Unsustainable **Oil**

The University of Alberta Press

Unsustainable **Oil**

Facts, Counterfacts and Fictions

JON GORDON

Published by

The University of Alberta Press
Ring House 2
Edmonton, Alberta, Canada T6G 2E1
www.uap.ualberta.ca

LIBRARY AND ARCHIVES CANADA
CATALOGUING IN PUBLICATION

Gordon, Jon, 1979–, author
Unsustainable oil : facts, counterfacts and fictions / Jon Gordon.

Includes bibliographical references and index.
Issued in print and electronic formats.

ISBN 978-1-77212-036-3 (paperback).—
ISBN 978-1-77212-098-1 (EPUB).—
ISBN 978-1-77212-099-8 (kindle).—
ISBN 978-1-77212-100-1 (PDF)

1. Ecocriticism. 2. Environmental protection in literature. 3. Oil sands industry—Alberta. I. Title.

PN98.E36G67 2015 809'.9336
C2015-906493-7
C2015-906494-5

First edition, first printing, 2015.
Printed and bound in Canada by Houghton Boston Printers, Saskatoon, Saskatchewan.
Copyediting and proofreading by Joanne Muzak.
Maps by Wendy Johnson.
Indexing by Adrian Mather.

The University of Alberta Press is committed to protecting our natural environment. As part of our efforts, this book is printed on Enviro Paper: it contains 100% post-consumer recycled fibres and is acid- and chlorine-free.

The University of Alberta Press gratefully acknowledges the support received for its publishing program from The Canada Council for the Arts, the Government of Canada, and the Government of Alberta through the Alberta Media Fund.

This book has been published with the help of a grant from the Canadian Federation for the Humanities and Social Sciences, through the Awards to Scholarly Publications Program, using funds provided by the Social Sciences and Humanities Research Council of Canada.

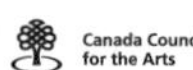

For Sam

Contents

Acknowledgements

THIS BOOK HAS BEEN EIGHT YEARS IN THE MAKING. To properly thank everyone who contributed to this reaching fruition is impossible in the space available. However, there are several people I need to thank by name. I want to thank everyone at the University of Alberta Press, especially Peter Midgley, Linda Cameron, Monika Igali, Cathie Crooks, Mary Lou Roy, and Alan Brownoff. For a careful reading of the manuscript that went far beyond copy editing, I would like to thank Joanne Muzak. For encouragement and insight along the way, I want to thank Paul Hjartarson, Betsy Sargent, Daniel Coleman, Warren Cariou, Diana Brydon, Imre Szeman, Sara Dorow, Dominique Perron, and Sheena Wilson. For a sense of belonging in an academic community, I wish to thank the members of the Association for Literature, the Environment, and Culture in Canada—especially Cate Sandilands, Pamela Banting, Rob Boschman, Susie O'Brien, Rita Wong, Jenny Kerber, Lisa Szabo-Jones, and Matthew Zantingh. To paraphrase William Stafford, the alchemy of your attention drew better writing from me than would otherwise have been possible. Credit for anything of value recorded here is, then, shared. I bear responsibility for any errors or deficiencies. For all kinds of support over the years, I am grateful to my family. Elizabeth, this would not have been possible without you. Samuel and Fredrick, you help sustain my hope for the future.

Portions of this book have been previously published, in different form, in *Countering Displacements: The Creativity and Resilience of Indigenous and Refugee-ed Peoples*; *Sustaining the West: Cultural Responses to Canadian Environments*; *Imaginations: Journal of Cross-Cultural Image Studies* 3.2 (2012); *The Goose* 12 (Winter/Summer 2013); and *American Book Review* 33.2 (2012). Many parts were initially conceived while writing conference proposals and first drafted as conference presentations. My thanks to the organizations at whose conferences those papers were given—the Modern Language Association (MLA); the Association for Literature, the Environment, and Culture in Canada (ALECC); the Association of Canadian College and University Teachers of English (ACCUTE), the Canadian Association for Commonwealth Literature and Language Studies (CACLALS), the Association for Commonwealth Literature and Language Studies (ACLALS), TransCanada, and Under Western Skies.

My thanks also to the many individuals, organizations, and institutions that gave permission for their work to be included in whole or part in this book. For permission to include images I want to thank the University of Alberta, David Dodge and the Pembina Institute, Greenpeace, Syncrude, the Government of Alberta, and the BC Teachers' Federation. For permission to quote text, I would like to thank John K. Samson for agreeing to let me quote from "Everything Must Go!"; Kaitlin Almack for allowing me to include "Thoughts of an Anthropologist"; Mike Gismondi, Debra J. Davidson, and Springer Press for allowing me to quote extensively from *Challenging Legitimacy at the Precipice of Energy Calamity*; J. Iribarne for "Cetology"; Kathryn Mockler for "Pipeline"; Mari-Lou Rowley for "In the Tar Sands, Going Down"; Melissa Sawatsky for "Say"; Allan Stoekl for *Bataille's Peak*; Colleen Thibaudeau's estate for "All My Nephews Have Gone to the Tar Sands"; Alison Watt and Pedlar Press for "Creekwalker" from the collection *Circadia*; and Rita Wong for "J28" and "night gift." Every effort has been made to acquire permission for copyrighted materials included in *Unsustainable Oil*.

Prologue

Fast Food Vacation

When Elizabeth got off the phone with her mom and asked if I wanted to go to Florida with her parents I said "No" immediately. "Absolutely not," may have been my exact words. I did not particularly want to go away over Reading Week. I did not want to go to Florida (Orlando, no less), specifically. I did not want to go away with my in-laws and spend a week in a time-shared resort condominium with them. I don't know that I ever really changed my mind...

In her book, *Bossypants*, Tina Fey provides "The Rules of Improvisation That Will Change Your Life and Reduce Belly Fat":

> The first rule of improvisation is AGREE. Always agree and SAY YES... Now, obviously in real life you're not always going to agree with everything everyone says. But the Rule of Agreement reminds you to "respect what

your partner has created" and to at least start from an open-minded place. Start with a YES and see where that takes you. (84)

I imagine taking a stand:

"I won't go."

"Why not?"

"I don't want to."

"Why not?"

"It's so far. How will Sam be on the plane? He's only one."

"He'll fly free."

"We have so many things we should do here during that time. I'm supposed to be writing about bitumen extraction and climate change. Flights are so expensive. Air travel is one of the worst things we can do for the environment."

"But how do I tell my parents that?"

"...I don't know."

There didn't seem to be a polite way to say "no."

In 2013 Suncor launched an advertising campaign titled "What Yes Can Do"; the television spot encouraged viewers to "come see what yes can do" by visiting www.whatyescando.com.

I went to see.

On the main web page they list:

Environment: Can we raise the bar on environmental performance?
Innovation: Can we be leaders in innovation?
Social: Can we help make communities stronger, resilient and sustainable?
Economic: Can we help sustain our quality of life and build our future?
People: Can we be concerned about the environment and work for an energy company?

I tried answering "no." I thought, do I have to honour what my partner has created? Do I get to choose my partner?

Sitting in the passport office waiting for my number to be called, I think about all of the people in all of the passport offices across the country and all of the places they are planning to travel and all of the oil necessary to make that travel possible. How much energy is required to fly a plane from Edmonton to Orlando?

It is a vacation neither of us really wants, but it's a generous invitation.

Suncor gives visitors to its website the chance to "Join the Conversation"—on a page titled "Talking about Yes"—where individuals can comment on threads like

1. "A recent Ipsos Reid survey shows that 80% of us believe that conversations about the oil sands should be based in science. What information do you want to hear about?"

Interestingly, that thread had zero comments when I visited. Maybe people feel unqualified to enter into such "conversations based in science." Maybe people recognize that "the scientific base" for the "conversation" is always mobilized to manufacture consent by linking scientific facts with an ideological narrative of a counterfactual future, and their silence is a way of withholding their consent. Maybe they recognize that a solid "scientific base" does not exist, that a lot of basic science has not been done. Maybe they recognize, as David Orr has it, oil "requires technologies that we are smart enough to build but not smart enough to use safely."

I thought: "Yes, and can you tell me why industry has not made an assessment of cumulative effects a requirement of new project approvals?"

In the pool with Sam at the resort it is idyllic: warm sun, fluffy clouds, splashing and giggling and tossing up in the air. "Swim to Mom, Sam." "Kick your legs."

I go to get towels. Elizabeth is holding him in the water and he's kicking his legs, swimming toward his Gram. I wish I had the camera so I could record it. He is so happy...

How many small children suffered to make that moment possible? How many barrels of oil did it take to get us there? How much blood was in that oil?

I tell myself our not going would not prevent that suffering, those deaths. Someone else would have filled our seats on the plane. We still would have been using oil at home, to heat our house (okay, so that's natural gas), drive our car. How can the suffering be avoided? Derrida says that choosing not to sacrifice one other always results in the sacrifice of other others. What do we do with the guilt this creates? How can we forgive each other? Ourselves?

Jamaica Kincaid writes about North Americans visiting Antigua:

> That the native does not like the tourist is not hard to explain. For every native of every place is a potential tourist, and every tourist is a native of somewhere. Every native everywhere lives a life of overwhelming and crushing banality and boredom and desperation and depression, and every deed, good and bad, is an attempt to forget this. Every native would like to find a way out...But some natives—most natives in the world—cannot go anywhere...They are too poor to escape the reality of their lives; and they are too poor to live properly in the place where they live, which is the very place you, the tourist, want to go—so when the natives see you, the tourist, they envy you,...they envy your ability to turn their own banality and boredom into a source of pleasure for yourself. (18)

George Grant: "In Nietzsche's conception of justice there are other human beings to whom nothing is due...Those of our fellows who stand in the way...can be exterminated or simply enslaved. There is nothing intrinsic in

all others that puts any limit on what we may do to them...Human beings are so unequal that to some of them no due is owed"
(*Technology and Justice* 94).

> The second rule of improvisation is not only to say yes, but YES, AND. You are supposed to agree and then add something of your own...To me YES, AND means don't be afraid to contribute. It's your responsibility to contribute. Always make sure you're adding something to the discussion. Your initiations are worthwhile. (Fey 84)

2. "Do you think that industry and corporate Canada have an important role to play in ensuring a good quality of life for the future?"
(1 comment)

I thought: "Yes, and I think that role will require stopping bitumen extraction, will require recognizing that the end of easy oil means the end of continuous growth."

In Orlando, at least in the part where we're staying, it seems as if the whole place exists solely for tourists: mile after mile of hotels, resorts, condominiums; restaurants, souvenirs, gas stations; Sea World, Disney, Epcot. The natives, it seems, if there are any, exist to serve the tourists.

3. "How can industry and individuals partner better on finding positive solutions for today's complex issues?" (1 comment)

I thought: "Industry can start by not asking loaded questions all the time, by working against the rhetoric of public relations, so that individuals don't always have to start with 'yes.'"

I keep hearing that line from The Weakerthans: "the outline to a complicated dream of dignity."

Dignity: The quality of being worthy of something; desert, merit. *Obs. rare.*

4. What is the role of industry regulators on environmental performance?" (4 comments)

I thought: "I'm no expert, but it's, maybe, to, umm, you know, regulate?"

"What are you going to have?"

"I think I'll get the Grouper."

"Isn't that an endangered fish?"

"Is it? The menu says they're committed to sustainable food supplies... at least it's local?"

5. "How can we work together to make positive contributions to communities? An Ipsos Reid public opinion poll shows that only 12% of Canadians give the industry an A ranking in this area." (1 comment)

I thought: "Broaden the definition of 'communities' to include human and non-human others."

The next rule is MAKE STATEMENTS. This is a positive way of saying, "Don't ask questions all the time." (Fey 85)

"So, how do you feel about this place now?"

"It's okay, I guess."

"You don't want to make it a yearly destination?"

"Absolutely not."

"What's your objection to it?"

"It reminds me of Las Vegas. It's like one of those made-up places, completely artificial. I imagine this all used to be swampland that was 'reclaimed' although I don't actually know that. But where are the mosquitoes? Where are the bugs? Surely a semi-tropical place should have bugs."

"But aren't there nice things about it?"

"Of course. I don't want to be chased inside by bugs all the time..."

"Are we just being pretentious in our disdain for places like this? Lots of people probably imagine this as the ideal holiday."

"I don't think it's just pretension. There are lots of things wrong with fancier holidays, but this is like a fast food vacation. Go to any of these resorts anywhere in the US, or probably the world, and you're going to get basically the same thing—a pool, a playground, a conference room where they'll try to get you to buy in to the time-share. Just like if you go to any McDonald's anywhere in the world you can get the same cheeseburger. I don't think it's just pretension to say that there are problems with McDonald's. That doesn't mean eating at fancy restaurants is a moral good..."

A dream of dignity is complicated by the ways in which our having dignity, or feeling that we have dignity, that we are worthy of something, is dependent on the sacrifice of the dignity of others or other others.

In the song the speaker's outline to his complicated dream is one of the things he's selling at the garage sale to pay his "heart's outstanding bills." He's willing to sell that dream of dignity for "a place where Awkward belongs, or a phone call from far away with a 'Hi, how are you today?' and a sign that recovery comes to the broken ones. Or best offer."

Maybe the best offer will turn out to be a week at a resort in someplace warm...not a place to belong but a place to get away to, not a sign that recovery comes but a break from the broken banality of an ordinary life. Accepting the offer of such an escape, though, depends on not asking certain questions.

For whom is such an escape impossible? What sacrifices are required, of us, of others, and of other others, to make such an escape possible?

6. "How do you define a sustainable community?" (11 comments)

I thought: "How do *you* define a sustainable community? Call me a cynic, but it seems like you're just fishing for ideas you can steal to put in your ads."

> Lastly, THERE ARE NO MISTAKES, only opportunities...In improv there are no mistakes, only beautiful happy accidents. And many of the world's greatest discoveries have been by accident. I mean, look at the Reese's Peanut Butter Cup, or Botox. (Fey 85)

George Grant: "Recent centuries have time and again witnessed one generation laying down the theoretical basis for such iniquities and leaving it to the next generation to take the words seriously and live by them. As common sense takes practice seriously and theory cavalierly, it has been popular to condemn those who commit the iniquities, but not those who justified them. Why should this be?...To write false theory may be more reprehensible than to organize a school badly, the bad organization of a school more harmful than robbing a bank" (*Philosophy* 87).

7. "How would you put your environmental ideas to work in an oil and gas job?" (2 comments)

I thought: "Sabotage?"

I only saw him for a moment, walking along the path towards Wakoola Springs beside Lake Opal, with his sunglasses and sandals, shorts, and a towel tossed over his shoulders. "Paradise" he said. I definitely heard the word "paradise." I thought, "hardly," but, if it's not this warm, sunny, bug-free place, then what is it?

8. "What's the role of the individual consumer in environmental sustainability?" (9 comments)

I thought: "Stop and think. Doubt. Question. Buy less. Don't always say 'yes.'"

The Orlando International Airport:

"Is this the line we're supposed to be in?"

"There's a wheelchair sign up there."

"Not everyone here is in a wheelchair."

"We have a stroller."

"Does that count?"

"I don't know."

"Look behind us."

"Well I don't want to change lines now."

"Who organized this place?"

"Why is it so busy on a Thursday morning?"

"Surely there are lots of people coming through here regularly; you'd think they could explain things better."

"This is ridiculous."

"Have we moved?"

"Barely."

"Air travel pretty much spoils any vacation. I don't think it's worth the hassle to have to go through this."

9. "How do you think innovative thinking will affect the future of the energy industry?" (4 comments)

I thought: "Hopefully by making it obsolete."

"I want you to understand the reason we can't go to Florida again this year, Sam. The cost to the planet is too high. Flying in an airplane—yes up in the

air, that's right—releases a lot of smoke which gets trapped in the air and then we breathe it in and it makes us sick; it makes us too hot. I know it's cold here in the winter, but it's not as cold as it used to be and other places are even warmer and we don't want to make other people too hot, do we? I know Gram and Gramps are still going and lots of other people will still go, but Mom and I decided that we shouldn't, and I know that's different than can't, but we believe it's important to think about the consequences of our actions and I know we went before but even if you do something wrong once it doesn't mean you should keep doing it. No that doesn't mean I think Gram and Gramps are wrong for going—everyone has to make their own decisions—but we'll see them when they get back. And...you...you can't build snowmen in Florida—do you want to go build a snowman?"

10. "Together, can we help make communities stronger, self-sufficient and sustainable?" (3 comments)

I thought: "Yes, and it requires you doing exactly what I say..."

I thought: "No, but separately we can fight long pointless battles over what is right, moral, pragmatic, realistic, honest, accurate, productive, progressive, sustainable, and scientific..."

I thought: "Yes, but only if this conversation starts with an acceptance that, in the end, those communities may say 'no' to bitumen extraction."

Was our trip to Florida a mistake? Was it a happy accident? What opportunities did it generate? Can those opportunities compensate for the sacrifices involved?

I'm sure I don't have the answers, but what follows is my attempt to think about the questions.

I accept the improvisational nature of life, but I refuse to say "yes" to every suggestion; sometimes we need to see what "no" can do.

Introduction

Unsustainable Rhetoric

taking a much needed pause for thought
on tarsands Highway 63
on the 401
on CN rail tracks
with Aamjiwnaang courage
a human river on Ambassador Bridge
time to stop & respect
remember we are all treaty people
unless we live on unceded lands
where rude guests can learn to be better ones
by repealing C45, for starters

—RITA WONG, "J28"

The essential requirement for change is that we change our minds about what we are here for, what we are fitted for as human beings and therefore what our stance or comportment should be inside modern technology and the empire where we find ourselves.

—ARTHUR DAVIS, "Did George Grant Change His Politics?"

TAKING A PAUSE FOR THOUGHT: what a simple and profound notion. Rita Wong's reading of the Idle No More movement, written in "J28," articulates what is urgently needed in our relationship to bitumen, and, more broadly, nationally and internationally, in our relationship to First Nations peoples and the lands on which they live, on which we live. Many people are working to rethink what those relationships should be, but, at the same time, the machines continue humming along and efforts to prevent, contain, co-opt, and manage the forms this rethinking can take also continue; the pause created through direct action is dismissed as *untoward* or turned against itself into the narrative of liberal progress.[1] We continue to be told that our stance towards the world should be one of doing, rather than thinking about what we are doing, why we are doing it, and how we might do things differently. This book tries to take up Wong's call in relation to bitumen extraction and explores the potential for literature to interrupt the relentless justifications and rationalizations *of* and *for* the status quo—that modern, liberal, technological society is free and autonomous and any limits we encounter are conditional, provisional, negotiable. It argues that fictional literature about bitumen extraction and its consequences can help readers rethink what our stance or comportment should be within the system where we find ourselves, and, further, that our stance should be untoward such a notion of progress: we should be difficult to manage, restrain, or control; difficult to manipulate, stubborn, and stiff. We should be marked by a lack of reason, when what passes for reason is so inadequate.

In a 2007 Alberta Treasury Branch (ATB) commercial, viewers were told that thinking "we'll never get the oil out of this sand...is not the attitude that built this province." The belief—that anything is possible, that challenges can and should be overcome—that did build it, the ad implies, is the product of technological modernity and serves to perpetuate it.[2] The view of bitumen as an untoward substance is counter to what "Alberta" stands for: untoward bitumen needs to be made to submit, to yield, to be turned toward the creation of a sustainable future for modern liberal capitalism. I would like to think that we could also harness this belief to change the narrative: anything is possible in Alberta, so why not leaving oil and gas in the ground? Why not a post-oil future of reduced energy consumption? But the story of technological modernity does not go that way. That narrative remains in the realm of the impossible, the unspeakable.[3]

Mike Gismondi and Debra Davidson, in *Challenging Legitimacy at the Precipice of Energy Calamity*, attempt to apply the "can-do" spirit of the ATB commercial to creating a post-oil economy in Alberta:

> This small provincial state *has* exhibited an extraordinary degree of autonomy in the past and perseverance in the face of miserable odds and repeated failures...Premier [Stelmach] takes every opportunity to remind listeners of the determined, innovative spirit of Albertans, and despite the many, many topics on which we wholeheartedly disagree...on this particular topic he is right. It was just such stubborn determination that, over the course of five (!) decades, allowed for the commercial-scale production of bitumen, decades during which private capital only played bit parts...But this very history, oddly enough, does provide a rather compelling counterpoint: if the Albertan state apparatus could so diligently see tar sands development to fruition despite the naysayers, it can see other challenging visions to fruition as well, including a Transition to a post-carbon economy. (219)

However, as they note, for such change to happen, the government would first need to be motivated to promote this transition, which raises the questions: Who motivated the development and who will motivate the transition, especially given the inertia of the status quo? Further, the five decades of commitment to unlocking the potential of the resource were, in many ways, the high point of the power of the nation-state and of cheap energy. Today, in 2015, that power is greatly reduced, with governments beholden to transnational corporate wealth (see Findlay).

Jeff Gailus describes how J. Howard Pew, of Sun Oil, invested $250 million in Great Canadian Oil Sands, now Suncor, following the five decades of "stubborn determination" by government—and the scientists, such as Karl Clark and Sidney Ells, they employed—that allowed commercial-scale production to get started: "Although Alberta's bitumen deposits were difficult and expensive to extract, [Pew] reasoned that they had the potential to provide the United States with an abundant source of oil far in to the future...He knew, or at least hoped, that bitumen's time would come. He need only toil and wait" (Gailus, *Little* 4). In this version, it is the stubborn determination, and capital, of the American oil tycoon that eventually gets the industry started. In Paul Chastko's history of the resource, *Developing Alberta's Oil Sands: From Karl Clark to Kyoto*, he describes the combined efforts of government and private capital in establishing the bitumen industry (xv). Regardless, these histories all show a motivation to secure the future by controlling an energy source that had been recognized since before the first European explorers saw it. Both the Alberta government and industrial capital were highly motivated to exploit this resource; they continue to be motivated to exploit it as conventional (read "cheap") oil becomes increasingly scarce around the world. And, as Natural Resources Minister Joe Oliver declared in April 2013, "either Canada finds a way to develop and sell its immense resources now, or opportunity will pass to others" (Cattaneo, "A Battle"). Provincial and federal governments still see bitumen as its "golden goose" (Lefsrud), even as this view becomes more difficult to maintain due to environmental damage and failure to deliver the promised economic results.

For Nietzsche, truth "*is the phenomenon of drawing or accepting the horizons within which one lives.* Making something true consists of making something *unquestioned*, in acknowledging it as a horizon" (Strong 50, emphasis original). The narrative of technological modernity—of change, progress, sustainability—*depends* on continued oil extraction and the burning of fossil fuels, *depends* on continued growth: this premise must remain unquestioned for the narrative to be true.[4] The logic works this way: if we take the money we get from using up this resource and destroying this ecosystem and endangering this watershed and these species, and we put that money into new technologies, renewable energy, and reclamation, then we can achieve "sustainability." This is a very powerful narrative, but *it is the same narrative that has led to the current situation.* It is the status quo, despite being cast as a new vision. Albertans such as Satya Das and Ken Chapman are seen as progressives on bitumen because they argue that extracting bitumen can give us the wealth to create a sustainable future.[5] This narrative is focused on meeting the increased energy demands that are projected for the future and making sure that *when the oil runs out* there will be new energy sources to take its place.

The notion that future technological advances, economic growth, and human progress will lead to the worldwide society of free and equal people begins from a premise of insufficiency: the world, as it both was and is, is inadequate for human needs; the key problem is scarcity. The idea that the world is somehow inadequate is not a new story. It is the narrative of deficiency that was told by Euro-American settlers of the New World (Kaye 5); it is the story of the Enlightenment as the universal and homogenous state of free and equal people that requires some groups to be remade in the image of others. As Frances W. Kaye argues, the deficiency paradigm implies "that Euro-North Americans were entitled to the land because they were Christians and because European-style commercial agriculture was a 'higher' use of the land than hunting and gathering and riverine horticulture" (61). This view problematizes Gismondi and Davidson's belief in Albertans' "can-do spirit"; the present situation, from which they argue we need to transition, is the result of the "can-do spirit" being imposed on a place, and peoples, that spirit viewed as fundamentally deficient. Can this same spirit solve the problems it created? Kaye argues that

> extraction of fossil fuels, and particularly oil and gas, has played a large part in the economic prosperity—and subsequent economic busts—of the region. Extraction comes with certain environmental degradations that emphasize the expendability of the place and its human and non-human residents. Alberta's oil sands are north of the Great Plains [which Kaye's study focuses on], but the vast expenditures of water, energy, and habitat in producing oil are resonant with the petroleum industry's history from Texas and Oklahoma up through Wyoming and the Dakotas to Alberta and Saskatchewan. Roger Epp's consideration of how they add to

> the "de-skilling" of the rural West is a twenty-first-century explication of the deficiency paradigm. (13)

This paradigm, which sees the rural as inferior, remains firmly entrenched despite ongoing forms of resistance. In "Continuing Dispossession: Clearances as a Literary and Philosophical Theme," Ian Angus argues both for the critical value in gaining distance from one's place, in order to understand it better, but also for the hope located in inhabiting a place *as one's own*. He writes,

> Empire is the dominance of the way of life of a people by an external power. We may define dispossession as being pushed out of one's place by an external power that makes impossible the continuation of the patterns of work and culture that have sustained life until that historical moment. Inhabitation, the negation of this negation, is the inverse. It is the sustenance of the life of a people that is expressed in work and culture through the continuity of time. After dispossession, such inhabitation becomes both the object of nostalgia for a pre-historic condition and that of desire for a future state. Inhabitation is the name for a mode of existence without empire. Thus, it may become the locus for hope. (10)

He goes on to note, "We have in Canada the example of peoples whose thinking is in place," citing James (Sákéj) Youngblood Henderson who "states that aboriginal knowledge 'reflects the complexity of a state of being within a certain ecology' and goes on to explain that 'experience is the way to determine personal gifts and patterns in ecology. Experiencing the realms is a personal necessity and forges an intimate relationship with the world.' Such thinking would be, in the terms that I have suggested here, the active component in inhabitation" (10). Such inhabitation has existed, and manages to persist, in the boreal region of Alberta and Saskatchewan, under which bitumen deposits lie, but it is precisely this inhabitation that is at risk due to the extraction that is occurring. Fort McMurray is a locus of imperial displacements.

For Angus, considering English Canada,

> Locative thought is the thinking of a people seeking its place and therefore of a people that has not yet found its place. A people that has undergone dispossession. In our case this is not a single clearance [as with the Highland Clearances] but a plurality which encounters fellow stories not in the site of the original dispossession but in the site after immigration structured by the hope, or dream, that one can own and keep one's own land. The irony is intended. The hope of keeping one's own land occurs in the place where dispossession has been visited upon the original inhabitants. ("Continuing" 10)

Such locative thought is happening in Fort McMurray, but it is by no means the predominant mode of thought, as Sara Dorow and Goze Dogu's study "The Spatial Distribution of Hope in and beyond Fort McMurray" shows. They write, "For many of the long-time residents, hope is embedded in and for the community of Fort McMurray itself" (273). We might think of these residents as engaged in "locative thought" in Northern Alberta. However, "for many in the mobile worker population, Fort McMurray is where hopes for other places, times, or people are enabled or frustrated; and for yet another set of people, namely, recently arrived professionals in the oil business, hope is flexibly able to put down roots in this place (or a number of places) for as long as is possible or prudent" (273). Their "locative thought" is directed, at least partially, elsewhere. This diversity of locative thought, which, in not including the Aboriginal population, is only partially representative of the region, shows some of the complexity of representing this place. The competing forms of inhabitation and dispossession that meet in the bituminous sands further this complexity: those whose "locative thought" is directed elsewhere may well justify continuing dispossession in and around Fort McMurray in the service of the place where they come from, where their families are, where their hope resides. Therefore, when Angus argues, "In seeking inhabitation, one of our key tasks is to recognize the inhabitation of others," we need to keep in mind the difficulty of doing so. He continues,

> It would be tempting to say that the plurality of stories of dispossession can be totalized into a universal concept of not-dispossession, emancipation through inhabitation, or justice...[However] it is not possible to say whether inhabitation in this universal sense is a political project capable of realization. But without such a horizon, the projects we imagine will not speak truthfully to our condition, but will degenerate into the apology and ideology of official culture. So we must leave the place of inhabitation indeterminate between a political vision of the good life and a philosophical image of justice. Locative thought needs hope and seeks it in "the modest trace elements of people's secrets held in common," as Myrna Kostash said, even though the place of this hope remains indeterminate—for we are not in place but on the way. ("Continuing" 10)

The indeterminacy of "being on the way" is both a hopeful and a frightening prospect. It entails both the possibility of reversing imperial dispossession and the possibility of its continued dominance.

Unsustainable Oil looks to literature for the possibility of inhabitation, for the alternative paradigm to imperial displacements and dispossession. Gismondi and Davidson write,

> as noted by Thomas Kuhn, a paradigm shift cannot occur until a viable alternative paradigm has emerged, and despite decades of dialogue on sustainable development, the precautionary principle, and simple living, all are replete with contradictions of their own. And most importantly, all the alternatives on the table also imply costs, not just to a small privileged elite, but also to material-consuming middle classes everywhere, not to mention a burgeoning global underclass that rightfully demands a significant *increase* in material consumption just to meet basic needs. (196)

I hope that in interrupting the dominant discourse, thereby creating a "pause for thought," a viable alternative paradigm can be created, or at least glimpsed. C.A. Bowers asks "whether the multiple dimensions of science...can lead to the development of an eco-justice pedagogy. Or is science propelling us further in the headlong rush toward the ecological catastrophe that will result from globalizing the Western consumer lifestyle?" (*Eco-Justice* 78). I take his recognition of the complexity of any answer as general agreement with Gismondi and Davidson's point, but also as a complication of it: all alternatives imply costs, but while the burgeoning global underclass can rightfully demand increases in material consumption, the justice of that demand is not a validation of Western culture. Our consumer lifestyle has eroded the community resilience of other cultures, making them vulnerable to the adoption of Western culture. We need, therefore, to "unlearn our privilege as our loss," in Gayatri Spivak's famous phrase ("Criticism" 164). The "costs" that Gismondi and Davidson identify, then, while undoubtedly pointing to the complexity of the problem, might better be recast as an opportunity for unlearning, to engage in locative thinking, and to move toward inhabitation of this place that English Canadian culture has, to this point, mainly related to through displacement—of the Aboriginal populations, of workers from elsewhere arriving to work in bitumen related jobs, of the land itself. For those living in places with access to cheap fossil fuels and "privileged" enough to benefit from that access, the contemporary paradigm shift can enable us to see the losses that have resulted, and continue to result, from that "privilege." *Unsustainable Oil* argues that our goal should be to view the world as sufficient in itself: "the human response to the Great Plains, until a few hundred years ago, was to use it, appreciate it, learn it, and manipulate it, but not to replace it or make drastic changes" (Kaye 4). The view of the boreal forest was similar and endured longer there than on the Prairies. Indeed, though this view is no longer dominant, it remains in important residual ways that can be recovered, reanimated, and rearticulated to resist the deficiency paradigm.

Ravelling and Unravelling Bituminous Narratives

On my first research trip to Fort McMurray in the summer of 2008, at the height of an oil boom, I went on a bus tour of the Suncor bitumen processing facility and attended *Simple Simon*, a play by Michael Beamish at the Canadian Natural Resources Ltd. stage of the interPLAY Festival. Both the tour and the play were performances addressing the ideas of sustainability and community in and around Fort McMurray, but the bus was full, the theatre, nearly empty. The tour guide worked hard to generate both a sense of awe around the size of the development and to assure her audience that reclamation was occurring successfully. For example, when we saw a deer alongside the road while we were at the Suncor separation facility, our guide stated, "deer and other wildlife like the extraction plant and mining site because there is no hunting allowed," suggesting obliquely that hunting threatened wildlife conservation while strip-mining promoted it.[6] The actors in *Simple Simon* also worked hard to persuade their considerably smaller audience of the psychological consequences of working at a bitumen mining site and being isolated from family and friends.

Beginning just prior to Donny McLeod's return home to Flatbush, Alberta after his week-long shift in a work camp, the play charts a couple's joyful reunion followed by the slow-motion breakdown of their relationship. While the tour narrates bitumen extraction as sustainable development, the play reveals the psychological sacrifices required of workers and their families on which development depends. It raises questions as to whether these personal sacrifices, and, by extension, social and ecological sacrifices, are sustainable.[7]

In "Metamorphoses of a Discipline: Rethinking Canadian Literature within Institutional Contexts," Diana Brydon adopts Roy Miki's Penelopean metaphor of "unravelling the nation" (13) as "a call to recognize that the creation of any imagined community is a continuous work in progress, involving making and unmaking, learning and unlearning, aiming not to fix boundaries but to encourage movements across them" (13). If we can apply, to this claim about the imagined community of the nation, Gérard Bouchard's view that collective imaginaries (442) operate at sub- and super-national levels as well, then we may have a useful framework for considering the continuous making and unmaking of communities happening in Fort McMurray and the Regional Municipality of Wood Buffalo. However, we need to keep in mind that this making and unmaking is occurring on multiple fronts by multiple constituencies simultaneously.

If we think of an ecosystem as a web of interactions, the process of extracting oil from the bituminous sands of Alberta's boreal forest can be read as an unravelling of one such web. To justify and compensate for this unravelling, a series of new webs are spun: through reclamation; through supporting cultural, recreational, and educational institutions; and through economic benefits to workers, governments, corporations, investors. This ravelling also occurs in narrative. *Unsustainable Oil* explores this simultaneous ravelling and unravelling, which is occurring on multiple levels in multiple media and multiple locations. Industry weaves stories to explain and justify its actions; governments weave stories to justify their actions and persuade voters; environmentalists weave stories in attempts to shift public opinion to slow or stop the unravelling of the ecology of the boreal forest; academics, pundits, journalists, and bureaucrats bring together facts, opinion, and emotion to argue about the future of the bituminous sands; creative writers invent different worlds and imagine different futures as a way of uncreating other possible worlds and futures. *Unsustainable Oil* examines this making and unmaking of narratives and how they impact a particular place in a particular eco-zone.

We can read the bitumen bus tour as part of this ravelling. As a joint production of the Government of Alberta; the two largest oil sands producers, Suncor and Syncrude; and Fort McMurray tourism, the tour addresses imagined communities at municipal, provincial, national, and cosmopolitan levels by emphasizing social, cultural, artistic as well as economic profits earned through the separation of bitumen from the sand, marshes, fens, and forest that compose the preceding ecosystem. And it highlights the effectiveness of the reclamation that is occurring. The play, by contrast, can be read as an unravelling of this myth of progress through profit, an interruption of it, focusing on the human, communal losses required to earn corporate profits. The concept of community can link the human and the non-human exchanges involved in bitumen extraction, if we think of it as consisting of both our human and non-human "relations" (see King, *All* ix and Angus, *Emergent* 84). One way of evaluating bitumen development may be to consider what forms of diversity are lost and what forms are gained through the life cycle of the oil extracted, but the webs of interaction are so complex that any attempts at "full cost accounting" are doomed to incompleteness. In *Simple Simon*, the activities that surround Donny's life are not very diverse (beer, television, ATVs, sex with Margi if he's not too tired and she's not too mad at him). By the end of the play, the last activity is no longer an option. There is less diversity to the community in which he lives and works.

In the play, Fort McMurray is a place Donny goes to, profits from, and gets out of again as quickly as possible, as he attempts to maintain his family and community elsewhere. Dorow and Dogu's research into ways residents of the region imagine the place they inhabit suggests that, for many newcomers at least, this view of the area as little more than a source of money is a common experience: "Fort McMurray figured into the hopes of industry workers predominantly as a place to make some money" (283). Wood Buffalo, at least for some, is an abstract space of profit-taking to which individuals come to get their piece of the action, while "home," the places where their memories, families, dreams, aspirations, and futures reside, is elsewhere. For many, the future of Fort McMurray includes neither their children nor themselves, as they plan to retire elsewhere. Partly, this is due to the lack of a long-term care facility in the area (Klinkenburg, "Health Care").[8]

In *Simple Simon*, the mentally challenged Simon idolizes his older brother Donny and tries to mediate between the feuding couple. Ultimately, he teaches Margi to take pleasure in the simple things in life, as he does, and, when she leaves Donny and heads to Vancouver, Simon goes with her. While this may be inadequate as a counternarrative to the myth of progress—and it is not hard to see how a trip to Vancouver is enabled by the petroleum economy whose consequences she is trying to escape—the play still unravels the industry narrative momentarily: the narrative that claims the sacrifices of human and non-human nature, of ducks and bogs and marriages and coaching your kid's hockey team are compensated for by the profits earned, or by the fact that oil from other places may be less desirable, or by underwriting the creation of literature—like *Simple Simon*—that bitumen makes possible. Donny, failing to learn the lesson Simon is trying to teach, is left alone facing the prospect of returning to a job he hates,[9] without the prospect of even an imagined community to which to retreat. The audience is left to ask what purpose his sacrifices have served. We have a moment to consider what forms of diversity, what forms of community, are lost and gained through this development, to question the myth Donny has been living in, to reflect on the myths we live within.

Bitumen as Sustainable Resource

The bituminous sands of Northern Alberta contain approximately 1.7 trillion barrels of oil, 176 billion of which are said to be recoverable with today's technology (Suncor Energy, *2009 Report*). This enormous resource has been, and continues to be, the subject of great debate, a debate over the human, social, and ecological challenges of commodifying it. *Sustainability* has become a central term in this debate. However, this term has been variously and divergently defined depending on the writer's focus. Sometimes it refers to sustaining growth or profits in the industry; sometimes it refers to sustaining quality of life for workers, their families, the communities of Fort McMurray and surrounding area. Sometimes it means sustaining the ecological integrity of the boreal ecosystem. Sometimes it means recreating an ecosystem capable of sustaining itself after the extraction is complete. These divergences indicate competing discourses and competing ideologies in the contemporary cultural moment. *Unsustainable Oil* examines these competing discourses, arguing that the dominant discourse understands sustainability as a question of being seen to take environmental concerns seriously so that oil can continue to be produced, so that the oil industry can be sustained.[10]

By contrast, literature about bitumen tends to suggest a counterdiscourse, which questions the narrative of sustainable growth. It helps make apparent the impossibility of sustainable extraction and use of nonrenewable resources, calls on readers to recognize their implication in the discourse that sustains oil, and allows them to dwell, temporarily, in that impossibility. The temporary dwelling outside of the status quo creates the potential for a "transvaluation of values" (Chisholm 572). As with Ellen Meloy's vision of the post-nuclear testing in the Nevada desert in *The Last Cheater's Waltz*—which, Dianne Chisholm shows, "deterritorializes the State's deterritorialization of the desert as void or as so 'empty' of natural and human resources as to be ideal for atomic bombing without consequence" (583)—most literature that represents the bituminous sands of Northern Alberta shows the region as having value beyond that of the oil that can be extracted from it. Rather than inadequate and empty, it is sufficient and full. But continuing the status quo of displacement and dispossession threatens this sufficiency.

The industrialization of the bituminous region of Northern Alberta is a manifestation of a cultural belief in limitless progress and endless expansion; it is a symptom of that culture. Many attempts to articulate alternatives are taken as affronts or insults to those invested in the status quo. As Canadian philosopher George Grant argues, "To write false theory may be more reprehensible than to organize a school badly, the bad organization of a school more harmful than robbing a bank" (*Philosophy* 87). This is the case because the consequences of believing false theory can infect whole cultures and impact the entire world, while robbing a bank is a single, isolated event. The consequences of the false theory of limitless expansion continue to impact our world in accountable and unaccountable ways.

As part of an advertising campaign titled "What's Next," in 2013 the University of Alberta ran full-page ads in *National Geographic*, *Maclean's*, and the *Vancouver Sun* that take the form of a letter from chemical and materials engineering professor Sirish Shah (see Figure 1); overwritten on the letter in large capitals is the message, "OILSANDS TECHNOLOGY WILL SAVE MILLIONS OF LIVES." This line marks an important shift in verb form from the story that originally ran on the university's website; although the content there is essentially the same as in the print ads, the headline reads, "Oil Sands Technology Could Save Millions of Lives." The shift from the conditional to the simple future tense marks a disturbing move. In its title—"What's Next"—the campaign implies the nature of the University of Alberta's cutting edge research and suggests, perhaps in response to provincial cuts to post-secondary education in the 2013 budget, the value of the university to society: the next big transformative ideas are being discovered here, the ads seem to say. Interestingly, a third version of the ad also exists, with the headline "Oil Sands Technology is Saving Millions of Lives." In this version, as is so often the case in representations of technological progress, the future—what's coming next—is said to be already here: the technology that "could save" or "will save" actually already "is saving" human lives, and not just a few lives but "millions" of them. Whether or not the technology has yet saved any lives is not explained in the letter, and perhaps this lack of present lifesaving prompted the shift to

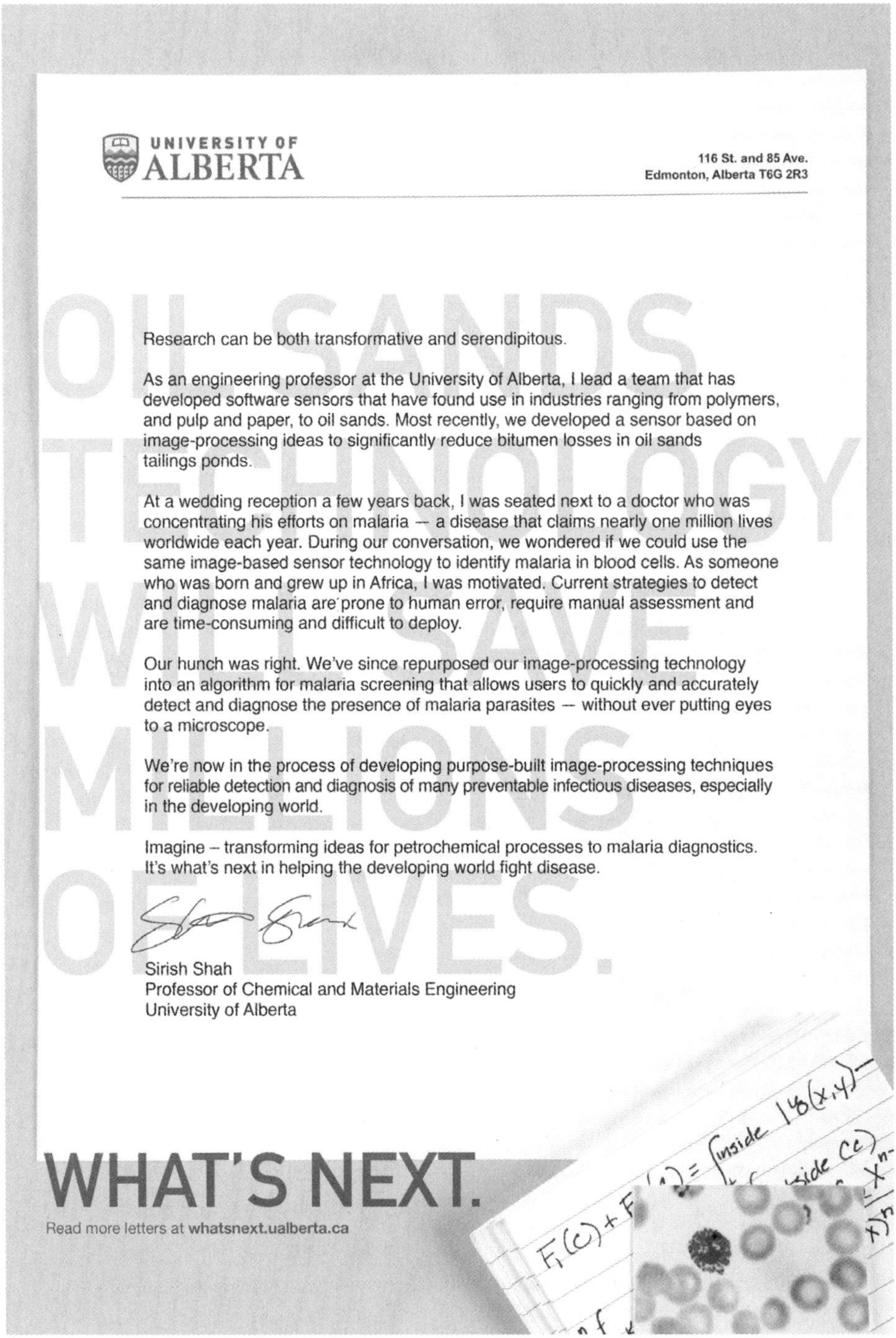

116 St. and 85 Ave.
Edmonton, Alberta T6G 2R3

OIL SANDS TECHNOLOGY WILL SAVE MILLIONS OF LIVES.

Research can be both transformative and serendipitous.

As an engineering professor at the University of Alberta, I lead a team that has developed software sensors that have found use in industries ranging from polymers, and pulp and paper, to oil sands. Most recently, we developed a sensor based on image-processing ideas to significantly reduce bitumen losses in oil sands tailings ponds.

At a wedding reception a few years back, I was seated next to a doctor who was concentrating his efforts on malaria — a disease that claims nearly one million lives worldwide each year. During our conversation, we wondered if we could use the same image-based sensor technology to identify malaria in blood cells. As someone who was born and grew up in Africa, I was motivated. Current strategies to detect and diagnose malaria are prone to human error, require manual assessment and are time-consuming and difficult to deploy.

Our hunch was right. We've since repurposed our image-processing technology into an algorithm for malaria screening that allows users to quickly and accurately detect and diagnose the presence of malaria parasites — without ever putting eyes to a microscope.

We're now in the process of developing purpose-built image-processing techniques for reliable detection and diagnosis of many preventable infectious diseases, especially in the developing world.

Imagine – transforming ideas for petrochemical processes to malaria diagnostics. It's what's next in helping the developing world fight disease.

Sirish Shah
Professor of Chemical and Materials Engineering
University of Alberta

WHAT'S NEXT.

Read more letters at **whatsnext.ualberta.ca**

FIGURE 1: "*What's Next: Oil Sands Technology Will Save Millions of Lives.*" [University of Alberta]

the future tense and to the promise of what will happen. Regardless, it is a significant leap to claim that improving diagnostic technology for a disease like malaria, which is the focus of the ad, will save lives given the developed world's long-term failure to invest in malaria treatment in developing countries and the high-cost of drugs to treat the disease. Perhaps most significant for my purposes here, though, is the claim that it is "oil sands technology" that will save lives, as if the technology somehow belongs to the oil sands. The letter from Dr. Shah makes it clear that the sensor technology has a wide range of applications, one of which is related to bitumen extraction. The linking of the technology to "oil sands" in particular seems intended to justify the University of Alberta's particularly close relationship with a controversial industry. Most troubling, however, is the move of taking a single discovery and linking it to a narrative of technological utopianism: what's next is a world where oil sands technology will save millions of lives. The consequences of such false theory are unaccountable.

Oil, then, is not the problem: a philosophy that presents oil extraction as the means to the good is the problem. Bitumen extraction is a symptom of a cultural narrative, and literature can intervene to change that narrative. For instance, literature could introduce this premise into the talk about sustainable community in Fort McMurray: the bituminous sands industry should end *before* all of the recoverable oil is extracted and that is what we should be putting the profits of the current extraction towards (not improving extraction, but ending it). Progress would not mean using less water or emitting less CO_2. It would mean producing less oil.

Tar, Oil, or Bitumen: None of It Is Sustainable

Eco-philosopher C.A. Bowers references Confucius's point that "if we are to rectify our relationships, we first need to rectify our language" (*Mindful* 5).[11] There is a continuing debate as to whether the bitumen deposits in Northern Alberta should be called "oil sands" or "tar sands" and the connotations of either choice distinguish positions in the political debate over their development. Andrew Nikiforuk, in *Tar Sands: Dirty Oil and the Future of a Continent*, makes his position clear in the title of his book but goes on to attempt to justify his language by criticizing the

Alberta government's logic that, "because oil is what is finally derived from bitumen" (Nikiforuk 12), it makes sense to call the deposits "oil sands." Nikiforuk responds, "If that lazy reasoning made sense, Canadians would call every tomato *ketchup* and every tree *lumber*" (12). He points to a distinction here, but this is a distinction that is not self-evident in the rationalized technocratic language of government. Indeed, our actions toward tomatoes and trees may often justify simply calling them ketchup and lumber. Nikiforuk's implied distinction between "tree" and "lumber" needs to be maintained, but we can no longer assume, as he seems to do, that it will simply be accepted as obvious. Gismondi and Davidson write on this point,

> Neither of the terms *tar* nor *oil* are technically accurate; the resource under northeast Alberta is what geologists refer to as bitumen. The increased use of the phrase *oil sands* is indicative of the conscious commitment by government to brand the industry in a more positive light in the face of its critics. And yet at the same time, the attention accorded to a name exemplifies the level of effort undertaken to assert control over a potential "runaway" discourse and is suggestive of the potential tenuousness of this state's legitimacy. (28)

Throughout *Unsustainable Oil* I have elected to use *bitumen* or *bituminous sands* to avoid the politically loaded tar/oil debate and to take a step towards rectifying our relationship to that substance by rectifying our language.

Nikiforuk's tree/lumber distinction also provides a starting point for thinking about how language use influences our actions. By considering the use of the term *sustainability* in relation to bitumen extraction, especially in relation to its fusion with the term *development* in government and industry documents, we can understand something of the complexity of rectifying our language as a first step toward rectifying our relationships. The phrase *sustainable development* serves a rhetorical function: confusing the distinction Nikiforuk wants to maintain. It yokes together two unrelated terms to imply a relationship between them. Graham Huggan and Helen Tiffin cite Wolfgang Sachs: "No development without sustainability; no sustainability without development...Development emerges rejuvenated from this liaison, the ailing concept gaining another lease on life" (31).

For critics such as Sachs, they argue, "sustainable development implies that it is economic growth, rather than the environment, that needs to be protected, and that environmental degradation is to be fought against principally because it impedes this growth" (32). On the other hand, Allan Stoekl makes the point, following Heidegger, that the reverse position also takes a fundamentally technological view:

> the sustainability proponents imagine a standing reserve that would somehow not deplete but rather conserve the resources that go into it... But nature would still consist of a reserve to be tapped and resources to be expended; the goal would still be the furthering of the stable human subject, the master of its domain...The quantified, mechanized destruction of Earth becomes the quantified, mechanized preservation of Earth. (*Bataille's Peak* 133)

Stoekl's reading of Bataille works to explore another energy, another economy, a heterogeneous energy, the cursed share, which is not in the service of humans: "The energy stored in and released from a strip-mined mound of coal is qualitatively different from, for example, the bodily energy discharged at the contact of an eroticized object" (138), or from reading a poem. In this way, literature participates in the heterogeneous economy, even though it also participates in the homogenous one (as commodity circulated for profit, material object requiring energy to produce). As Jean-Luc Nancy states, literature creates a "circulation [that] goes in all directions at once" (*Being* 3), even if some of those circulations are captured by the market. The multiplicity of these circulations are distinct from the controlled circulations sought by rhetoric.

Claims about the "sustainability" of bitumen extraction depend on what definition of sustainability is being used. The failure of all sides in the debate to recognize the multiple meanings of this central term has created polarized and entrenched rhetorical positions. The social and cultural values at play in bitumen development are often managed through the phrase *sustainable development*, a phrase whose two terms seem at odds unless what we are trying to sustain is development itself, in which case the adjective *sustainable* is superfluous or simply added for rhetorical purposes. If, however, on one hand, we want to sustain, in the sense of conserve, the natural diversity of the earth, and, on the other hand, to develop a more just, equitable, and prosperous civilization, then *sustainable development* functions to suggest that we can do both at the same time.[12] While this may be possible, it is not possible to gauge our success or failure unless we know what each term means, unless there is a value system guiding our actions, which, at the very least, brings to explicit discussion what we are trying to sustain and what we are trying to develop: do we want trees or lumber? Or, better, do we want forests or lumber?

In "Canada's Oil Sands," a publication of the Canadian Centre for Energy Information, Robert Bott uses language similar to that of the Alberta government when he writes, "The future of this resource will be decided in the laboratory" (38). My contention is that this assertion depends on a hidden premise: it must first be decided within a paradigm that justifies (and pays for) laboratory research. This foundational, unspoken, and often, it seems, un-thought assumption needs to be foregrounded as a first step to any substantive debate on bitumen extraction.

The work of George Grant, whose thought is characterized by both a profound commitment to human equality *and* to the value of tradition, can be helpful in this process. As much of my engagement with his thought in this book shows, his philosophy is also ecological in its consideration of the non- human other as good for its own sake. As such, he argues against the belief that "we have knowledge when we represent anything to ourselves as object, and question it, so that it will give us its reasons...These procedures are what we call in English 'experimental research'" (*Technology and Justice* 36). Bitumen, rather than being just an end product of technical processes, is a by-product of the culture that enables its production, a culture that privileges objective reason, a culture that can't see the forest for the lumber. Huggan and Tiffin summarize Val Plumwood's point that "the dualistic thinking that continues to structure human attitudes to the environment" is linked "to the masculinist, 'reason-centred culture' that once helped secure and sustain European imperial dominance, but now proves ruinous in the face of mass extinction and the fast-approaching 'biophysical limits of the planet'" (4). It is "those forms of instrumental reason that view nature and the animal 'other' as being either external to human needs, and thus effectively dispensable, or as being in permanent service to them, and thus an endlessly replenish-able resource" (4) that need to be dismantled. Bitumen's production depends on justifications of technical advances as being in the common good. The meaning of *sustainability* must be analyzed if we are to question the bases underlying a common sense faith in a future decided in laboratories.

To justify such a future, sustainability has been fused with the rationale of neoliberalism, repackaging the logic of technical advance and expansion as an answer to environmental criticisms. In 2008, the UK Advertising Standards Authority raised questions about such claims when it ruled, "in response to a [World Wildlife Fund] complaint, that a Shell advertisement touting its Canadian oilsands project as 'sustainable' was misleading and could not be published again" (Boswell and Butterfield A1). Jeff Gailus, in *Little Black Lies: Corporate and Political Spin in the Global War for Oil,* quotes University of British Columbia philosopher Alan Richardson, who calls this type of advertising "performative bullshit" since "Claims...that... activities are or will become sustainable '[have] the form of a commitment, but it is not a real commitment.' Corporations and governments know they have not made a commitment, but they act as if one were in effect in order to deflect public concern about future impacts" (71). The controversy around the Cumulative Effects Management Association (CEMA) targets the same lack of commitment. Established in 2000 to study cumulative environmental effects of bitumen extraction in Northern Alberta, CEMA has produced many recommendations over the years but has not created regulations or made cumulative effects a key factor in the granting of new licences to develop bitumen (Acuña). "Various groups have left CEMA, citing its ineffectuality (Severson-Baker, Grant, and Dyer). In 2013 the organization, which receives its funding from industry, saw that funding cut in half (S. Pratt, "Funding"), only to have it restored in early 2014 for a one-year period with certain conditions (McDermott). Mark Hebblewhite points out that

> energy development can affect almost all natural resources, including surface and subsurface hydrological processes, natural disturbance regimes such as fire, soil erosion processes, wildlife habitat, and wildlife

population dynamics...but mitigation is typically implemented on a site-by-site basis...[even though] impacts of development are most often felt through cumulative effects of not just one well site at a time but across large landscape scales on the order of thousands of square kilometers. (71)

The scale of the bitumen deposit in Canada's boreal forest emphasizes Hebblewhite's point even further.

At the same time as public concerns are deflected through advertising and misled through empty claims to assess cumulative effects, "productivity" has become the definition of sustainability through the government's use of the "Land Capability Classification for Forest Ecosystems in the Oil Sands" to evaluate reclamation.[13] Within this system, reclamation "does not mean 'tree-by-tree' restoration, but rather that the region as a whole should form an ecosystem at least as healthy and *productive* as what existed before development" (Bott 31, emphasis added). This discourse of productivity asserts that the land will be more humanly useful, more profitable, because its productivity will be oriented to marketable ends. Planting cereal crops and trees, raising bison for profit, and creating recreation areas constitute "reclamation." As Jennifer Grant, Simon Dyer, and Dan Woynillowicz, writing for the Pembina Institute, argue, the focus of reclamation is primarily "on commercial forestry" (21), dictating "the reclaimed target landscape—a forested ecosystem" (22). Replacing wetlands with forests can then be construed as an "improvement" to the land, creating, in their words, "a perverse situation where oil sands proponents claim there will be an improvement in land capability after reclamation" (22). The land will produce more "merchantable timber" (Grant, Dyer, and Woynillowicz 5) and is, therefore, better than what preceded mining operations. The amount of profit land can generate determines its value rather than the diversity of life it supports, even if the latter must be sacrificed for the former.[14] Trees are not trees, but, rather, lumber, or, more likely, pulp, the marketable commodity they are able to produce. The narrative of "sustainability" rhetorically mystifies this sacrifice.

The rhetoric around the use of "reclaimed" sites for the people of Alberta is perhaps even more significant than any potential profits to be made after the bitumen is extracted. As Nathan Vanderklippe writes in the *Globe and Mail,* the mineable bitumen area "could one day be Alberta's very own Lake District, a recreational haven complete with campgrounds, boating, fishing—even swimming." The twenty-nine proposed end pit lakes (EPLS), which will have fresh water pumped over process-affected water and tailings, are seen as part of the solution to the ongoing challenge of reclaiming tailings and as a much cheaper alternative to filling mined-out pits with dirt. However, as Vanderklippe notes, "It may take a century to find out what is left around Fort McMurray." The industry gamble is that the process will result in clean water; the government bet seems to be on the public accepting these lakes in place of the ecosystem that preceded them.[15] A CEMA "Backgrounder" on end pit lakes states, "Returning mineable oil sands developments in the Athabasca oil sands region to a state functionally equivalent to the natural conditions that characterize the boreal forest of Northern Alberta will involve a sophisticated suite of tools, techniques, and expertise. The end pit lake is an integral element of this task." Alberta Environment's website states, "CEMA's Reclamation Working Group has a mandate to provide recommendations to government to ensure that reclaimed landscapes within the region meet regulatory requirements, satisfy the needs and values of stakeholders, and are environmentally sustainable." This list connects three key elements that are often conflated in the use of the word *sustainable* or the phrase *sustainable development* but need to be considered separately. First, there are the technical-scientific regulatory requirements, which, regarding end pit lakes, industry is betting on its ability to meet. Second, there are the needs and values of stakeholders, a term that is often used to restrict whose needs and values count in terms of public consultations, but, given the timeline of reclamation, should include future Albertans, especially future generations of First Nations who may want to continue or reclaim traditions developed through thousands of years of connection to the region.[16] The third element is environmental sustainability, which needs to include not only the potential for the "reclaimed" ecosystem to be self-regulating after reclamation is complete but also for the biodiversity that pre-existed mining to return; this environment also needs to be capable of sustaining human life. These three elements have complex interactions in much of the rhetoric surrounding bitumen. The main focus of *Unsustainable Oil* is on the second item—the needs and values of stakeholders—and whether or not the public will accept the condition of the land that was leased to industry to extract the bitumen.

The *Oxford English Dictionary* gives the following definitions for *sustainability*: "Of, relating to, or designating forms of human economic activity and culture that do not lead to environmental degradation, esp. avoiding the long-term depletion of natural resources"; "Capable of being borne or endured; supportable, bearable. *Obs. rare*"; "Capable of being upheld or defended; maintainable"; and "Capable of being maintained at a certain rate or level" (which gives us the term *sustainable growth*). It should be obvious that developing the bituminous sands is not sustainable in the sense of "not lead[ing] to environmental degradation, esp. avoiding the long-term depletion of natural resources." Every barrel of bitumen produced both depletes this finite resource and increases the environmental degradation caused by extraction. Further, as the depletion continues, the degradation is likely to increase. Nikiforuk quotes Bruce Peachey of New Paradigm Engineering on the industry's water use: "We are presently mining the best ores. But as clay content increases, the volume of water needed in production will increase" (*Tar Sands* 62). Gismondi and Davidson elaborate on this point,

> ecomodernists tend to conveniently gloss over the many forms of environmental degradation that are caused by our economic activities, including energy extraction, presuming such consequences can be met with the same technological innovation that will enable the energy to continue to flow. [However]...the ecological impacts of resource production...*increase* in intensity as the quality of the ores declines. The ecological impact of producing one barrel of bitumen is far greater than the ecological impacts of one barrel of conventional crude, and the ecological impacts of the one-billionth barrel of bitumen will inevitably be greater than the ecological impacts of the first barrel produced. (147)

They go on to note that, "Tar sands EROI [Energy Return on Investment] estimates range b/w 4:1 to 3:1. When placed in context of net EROI, the actual potential of the Athabasca tar sands is far less attractive. The current EROI will not be maintained, furthermore, but will follow a continuous declining trajectory, as the quality of the remaining resource diminishes (requiring ever deeper drilling shafts, greater amounts of solvent, etc.)" (149). When the ratio hits 1:1, there will be no benefit in extracting more bitumen. There will be lots of bitumen left, but it will be no use to humans.[17] Until we get to that point, bitumen extraction means resource depletion and cannot be sustainable in the first sense.

The other definitions, though, offer more rhetorical wiggle room. First, is the development capable of being endured? By whom? For how long? With what forms of compensation? This is where we get into the type of contractual thinking—of compromise and trade-offs—that leads to royalty regimes, and arguments about the local nature of environmental damage and the global nature of profits. Nikiforuk quotes Charlie Fisher, former CEO of Nexen, disputing the findings of the royalty review panel: Fisher says, "When we talk about fair share, I don't know what that is" (*Tar Sands* 146). I have to agree, but I'm sure from a different perspective: What is the fair human share? What is the fair non-human share? How can these be separated? Analyzed? Given a dollar equivalent?

Second, is the development capable of being upheld or defended? Certainly, government and industry have managed to sustain development by this definition so far. Despite public concern over the ecological consequences, despite lawsuits and protests, former Alberta Premier Ed Stelmach (2006–2011) won a massive majority in 2008 with his "don't touch the brake" development policy and this policy was affirmed by Premier Alison Redford's (2011–2014) efforts to promote the building of the Northern Gateway and Keystone XL pipelines.[18] The choice of Jim Prentice as leader of the Progressive Conservatives in September 2014, and, thus, as premier of Alberta, did not result in any significant shift in energy policy. However, in May 2015, the nearly forty-four year PC rule of Alberta ended with the election of an NDP government under Rachel Notley, which campaigned on (among other things) promises of higher corporate taxes, a royalty review of the energy industry, and changes to carbon regulations. The significance of this change of government remains to be seen, of course; however, it may suggest that the PC's version of development is no longer capable of being defended. At the same time, Notley's new government has, since taking office, worked hard to reassure the industry, and markets, that it is committed to working with them to maintain a competitive energy sector.[19] Alberta's prosperity (and, indeed, Canada's) is said to depend on continued expansion of bitumen extraction industries.

Third, is it capable of being maintained at a certain rate or level? It seems to me that this is the sense in which sustainable development is most often used. Can we maintain the status quo—that is, growth? Can we continue to develop, in the sense of expand, the industry? Here it seems we have a catch-22. Our economic system is posited on continuous growth. Our planet is finite. While it may no longer be fashionable—given the glut of unconventional oil and gas hitting the market—to quote Marion King Hubbert, the Shell geophysicist who first theorized "peak oil," his words are pertinent:

> Our principal impediments at present are neither lack of energy or material resources nor of essential physical and biological knowledge. Our principal constraints are *cultural*. During the last two centuries we have known nothing but exponential growth and in parallel we have evolved what amounts to an exponential growth culture, a culture so heavily dependent upon the continuance of exponential growth for its stability that it is incapable of reckoning with problems of non growth. (qtd. in Nikiforuk, *Tar Sands* 172, emphasis added)

In *Living Oil*, Stephanie LeMenager begins by noting "the problem isn't that we're running out of oil, but that we're not" (3). She goes on to show how the shift from the "easy oil" of the twentieth century to the "tough oil" that's left in the twenty-first enables us to start to see how "energy systems are shot through with largely unexamined cultural values, with ethical and ecological consequences" (4). This shift creates the conditions for the battle over the future that Žižek describes ("Use"). So, though the Marxist literary critic Fredric Jameson has famously stated, "it seems easier for us today to imagine the thoroughgoing deterioration of the earth and of nature than the breakdown of late capitalism" (qtd. in Szeman, "System" 820), the cultural assumptions that enable that deterioration can be examined, questioned, and changed. However, government responses to the 2008 economic crisis and its aftermath seem to support these claims: mortgaging the future to get the economy growing again was, at the time, continually presented as unarguable, commonsense wisdom.[20] Thus, it seems, that, unless we have a profound *cultural* shift, development is what will be sustained until we reach a state of depletion. As William Marsden has it, we will remain "stupid to the last drop" of oil, or the last drop that can be extracted at a profit anyway.

Sustainable development is, then, for government and industry at least, primarily a way of turning trees into lumber, tar into oil, and critique into consent; a way to defend the status quo of growth at any cost. The phrase is used in the economic sense,[21] not the ecological. One step towards rectifying our language on the way to rectifying our relationships would be to seize this moment of economic turmoil and consign *sustainable development,* as George Orwell suggests doing with such "bits of verbal refuse," to "the dustbin where it belongs" (140). When *sustainable development* stops being an organizing term, a way of suggesting we can have our trees and lumber too, then we can really begin to talk about what our relationship to the bituminous sands of the Athabasca region should be. There is no such thing as sustainable development of a non-renewable resource.

Literature's Counterfactual Response to All the Factual Bullshit

David J. Tietge, in "Rhetoric Is Not Bullshit," points out that "the common perception of rhetoric as being a mode of discourse lacking substance, of being the epitome of empty embellishment, is prevalent in popular and political representations of it, as evidenced in its frequent appearance in phrases like 'once one gets past the rhetoric' or 'all rhetoric aside'" (230). *Unsustainable Oil*—while arguing that literature does have the capacity to suggest that which is beyond rhetoric, beyond words—is also engaged in *rhetorica docens,* "the theoretical treatment of words used to discover how language means among different agents, motives, cultural and social idiosyncrasies, and external events" (Tietge 231). At the same time, though, declaring that "rhetoric is not bullshit" imposes one meaning of the term *rhetoric, rhetorica docens,* in place of another, *rhetorica utens.*

The same word, *rhetoric*, is used both for the practice of using language to make things happen in the world, and for the study of that use. If we understand *rhetoric* in this way, as having these multiple meanings simultaneously, then it both is and is not bullshit: sometimes language *is* lacking substance or *is* just empty embellishment or *is* indifferent to the truth (as Harry Frankfurt argues in "On Bullshit"). The phrase *sustainable development* is rhetorical—it seeks to make something happen in the world—and it is, often if not always, bullshit. At the same time, it is also possible to use language to examine, expose, and understand such uses of language, as Tietge points out. My exploration of the phrase *sustainable development* here is also rhetorical in this sense. This is not to suggest that all instances of *rhetorica utens* are "bullshit" (or that all instances of *rhetorica docens* are not). I agree with American literary theorist Kenneth Burke that

> In pure identification there would be no strife. Likewise, there would be no strife in absolute separateness, since opponents can join battle only through a mediatory ground that makes their communication possible, thus providing the first condition necessary for their interchange of blows. But put identification and division ambiguously together, so that you cannot know for certain just where one ends and the other begins, and you have the characteristic invitation to rhetoric...When two men collaborate in an enterprise to which they contribute different kinds of services and from which they derive different amounts and kinds of profit, who is to say, once and for all, just where "cooperation" ends and one partner's "exploitation" of the other begins? (25)

In such a situation, and the extraction of bitumen is surely one such, it is unclear, and, ultimately, unknowable, what is bullshit and what is not. While, as I argue in Chapter 3, this situation can lead to an apathy that enables the status quo to reproduce itself, it is incumbent on those of us who have the opportunity to take a "pause for thought" to do so and to push against this unknowability and to investigate the ambiguous line between co-operation and exploitation.

Nonetheless, the emphasis on facts and reason continues to be reiterated in the debate over bitumen. On March 29, 2013, the *Edmonton*

Journal reported Colorado Governor John Hickenlooper's call to "bring the debate [over the Keystone XL Pipeline] back to fact-based discussions" (Graveland A10). Earlier that month, on Sunday, March 13, 2013, in a half-page, thirty-thousand-dollar ad in the *New York Times*, the Alberta government declared, "Keystone XL: The Choice of Reason." The ad stated, "America's desire to effectively balance strong environmental policy, clean technology development, energy security and plentiful job opportunities for the middle class and returning war veterans mirrors that of the people of Alberta" (D. Bennett A7). Therefore, it concluded, "choosing to approve Keystone XL and oil from a neighbour, ally, friend, and responsible energy developer is the choice of reason" (D. Bennett A7). Regarding this ad, a spokesperson for Premier Alison Redford explained, "It's important for Alberta to get the facts on the table as widely as possible" (D. Bennett A7). This focus on "facts" ignores Burke's point about the ambiguity of co-operation and manipulation: when one party marshals a set of facts to promote a co-operative endeavour, when does it become manipulation? Arguing from "facts" does, though, continue to resonate with "social or cultural values or norms" (Gismondi and Davidson 22) by suggesting that we live in a rational world that we can control.

While I am not making an argument against rational inquiry and argument, I do believe that we need to recognize, as George Steiner argues in *After Babel*, that

> the communication of information, of ostensive and verifiable "facts," constitutes only one part, and perhaps a secondary part, of human discourse. The potentials of fiction, of counterfactuality, of undecidable futurity profoundly characterize both the origins and the nature of speech...They determine the unique, often ambiguous tenor of human consciousness and make the relations of that consciousness to "reality" creative. Through language, so much of which is focused inward to our private selves, we reject the empirical inevitability of the world. (497)

Of course, this rejection is not only performed by one side in a debate.[22] Steiner's point is that this is a human characteristic. In *Slow Violence*, Rob Nixon asks, "How can we convert into image and narrative the disasters that are slow moving and long in the making? How can we turn the long emergencies of slow violence into stories dramatic enough to rouse public sentiment and warrant political intervention?" (3). *Unsustainable Oil* argues that the literary counterfactual is one such way. When considering the possible futures of bitumen, we should, then, look not just at the facts marshalled by the various sides, but also at the related and competing counterfactuals. This is the work that *Unsustainable Oil* undertakes.

Although the critical movement for the last thirty years or longer has been away from seeing literature as a privileged form of discourse, and counterfactuals are both a fundamental characteristic of human thought and used as a methodological tool in disciplines from physics to psychology to emergency preparedness, the ways in which literature explores counterfactuals are unique. Hilary P. Dannenberg, in *Coincidence and Counterfactuality: Plotting Time and Space in Narrative Fiction*, notes,

> People tend to construct counterfactual narratives in the everyday course of events. These can range from minor counterfactual responses formulated when the daily schedule goes wrong ("if I had not gotten stuck in a traffic jam, I would have arrived on time" [Fearon 1996, 50–51]) to the counterfactuals generated, often at times of personal crisis, to review one's life trajectory and formulate long-term regrets about missed opportunities (Gilovich and Medvec 1994; Kahneman 1995). Such alternate life scenarios are an everyday manifestation of the human urge for narrative liberation from the real world...Human beings' tendency to construct alternate worlds to the one they live in already begins with such natural counterfactual narratives. (110)

In the chapters that follow, we will see such counterfactuals played out in pro- and anti-bitumen narratives and in literary texts. Given the focus on the *factual* in both the pro- and anti-bitumen rhetoric, the *counterfactual* becomes an important site for both debunking claims to building a future based in reason and assembling alternatives to such futures. As Ray Bradbury once said, "The function of science fiction is not only to predict the future...but to prevent it" (qtd. in Sturgeon viii). Although not strictly science fiction, many of the literary engagements with Alberta's bituminous sands examined in *Unsustainable Oil* can be read as attempts, through counterfactual insertions into the discourse of "facts," to prevent the future promised by champions of bitumen extraction.

Another way of thinking through the distinction is through what philosopher John Caputo calls "radical hermeneutics." He writes,

> it is the claim of radical hermeneutics that we get the best results by yielding to the difficulty in "reason," "ethics," and "faith," not by trying to cover it over. Once we stop trying to prop up our beliefs, practices, and institutions on the metaphysics of presence, once we give up the idea that they are endowed with some sort of facile transparency, we find that they are not washed away but liberated, albeit in a way which makes the guardians of Being and presence nervous. Far from abandoning us to the wolves, radical hermeneutics issues in far more reasonable and indeed less dangerous ideas of reason, ethics, and faith than those that metaphysics has been peddling for some time now. Curiously enough, the metaphysical desire to make things safe and secure has become consummately dangerous. (7)

The dangerous desire to represent things as safe and secure operates through "facts" but also relies on linking those facts with counterfacts. We are told, for example, "our scientific expertise, which has enabled us to release the energy locked in bituminous sands, will also enable us to reclaim this land and guarantee a prosperous future for all." The first part of such a claim is factual; the second is not. The shift Caputo describes—yielding to difficulty—can be seen as analogous to the shift from "matters of fact" to "matters of concern" that Bruno Latour calls for, and which is discussed in Chapter 3. As Caputo suggests, such a shift can be frightening, as it is often more appealing to choose the simple over the difficult.

Dannenberg distinguishes between "upward" counterfactuals, which construct "a better consequent, improving on circumstances in the factual world" (119) and "downward" counterfactuals, which "construct a worse version of the factual world" (119). The Alberta government's ad, and much of the government and industry discourse around bitumen, might usefully be read as "upward counterfactuals": they imagine a better outcome in the future. Most literary representations of bitumen construct "downward counterfactuals": the future created by bitumen extraction will be worse than the present. While it is possible for both upward and downward counterfactuals to be equally difficult, in practice the upward narrative of infinite progress, in its simplicity, is far more dangerous than the downward narrative of limit and decline because it is so tempting and persuasive.

In "With an Eye Toward the Future: The Impact of Counterfactual Thinking on Affect, Attitudes, and Behavior," Faith Gleicher and her co-authors claim that "upward and additive counterfactuals are more likely than downward or subtractive counterfactuals to lead to plans for future behavior" (301). Gleicher and colleagues go on to argue, "Because of the power of counterfactual simulations to stimulate emotional responses and because of their inherent process component, they may be particularly effective in spurring individuals to engage in further simulation that directly influences behavior" (302). It seems to me, therefore, that the upward counterfactual narratives are on the side of the status quo and thus are more likely to influence behaviour. Arguing from "the facts" that continued petroleum extraction will lead to economic growth and prosperity for all offers a more appealing counterfactual than arguing from "the facts" that the future promised through bitumen extraction will be bleak. This leads to a widespread culture of denial. The upward counterfactuals, which present the world as simple, secure, and safe, are, as Caputo states, dangerous, and increasingly so in the era of climate change. The downward literary counterfactuals discussed in *Unsustainable Oil* do the work of "radical hermeneutics" that Caputo calls for, of restoring the difficulty to life. However, these texts leave us with the challenge of thinking through the question of persuasion, of rhetoric, of upward counterfactuals, to displace the hegemonic narrative surrounding bitumen development.

After years of training and practice in critical debunking, I am working to accept Latour's assertion that "the critic is not the one who debunks, but the one who assembles" ("Why Has Critique" 246). Such acts of assemblage may offer some of the hope of upward counterfactuals while also restoring the difficulty to life. At the same time, it seems that when it comes to the debate around bitumen development, there is a great deal of worthwhile debunking to be done. Therefore, *Unsustainable Oil* considers Latour's notion of debunking and assembly as a dyad, one perhaps analogous to Paul Ricoeur's hermeneutics of suspicion and affirmation which, for Ricoeur, "form a tension" rather than one having value "to the exclusion of the other" (Coleman 32). This book does not abandon that critical, debunking, suspicious intelligence completely but tries to recognize its limits and not to ignore the assembling, affirmative intelligence. Perhaps, then, the critic should be one who debunks *and* assembles. In *Vibrant Matter: A Political Ecology of Things*, Jane Bennett states that, though the practice of demystification is important, it is also limited. She notes, "A relentless approach toward demystification works against the possibility of positive formulations" (xv), and "demystification tends to screen from view the vitality of matter and to reduce *political* agency to *human* agency" (xv). With Bennett, I try to resist this tendency in *Unsustainable Oil*, at least sometimes. As she writes, "The capacity to detect the presence of impersonal affect requires that one is caught up in it" (xv), and my readings of literary texts try to retain this possibility. These issues are discussed in more detail in Chapters 3 and 4, but, in short, *Unsustainable Oil* argues that the discourse around bitumen is so polarized that the mediatory ground of rhetoric that Burke references is increasingly rare and that literature offers a space where entrenched positions can shift, where questions of co-operation and exploitation can be rethought, because literature is something people can get caught up in.

Each of the subsequent chapters puts one or more literary texts into conversation with government, industry, and media discourse about Alberta's bituminous sands. Throughout, the spectre of the 1,600 migrating ducks that died on Syncrude's Aurora tailings pond on April 29, 2008 haunt the subtext of the discussion, surfacing repeatedly as a reminder that existence is "unsacrificeable" (Nancy, *Being* 25). Dannenberg notes,

> research has found that people are more inclined to mentally mutate exceptional events than routine ones: "The more abrupt or discontinuous the change, the more likely people are to notice it, to try to explain it, and generate counterfactual scenarios in which they 'mutate' the departure from normality to the more customary and expected default value"

> (Tetlock and Belkin 1996 b, 33; see also Kahneman and Miller 1986). Counterfactuals are therefore often generated in dramatic or exceptional situations, and traumatic or negative outcomes in particular tend to trigger counterfactuals. (111)

On the one hand, this may suggest a kind of evolutionary response to trauma: in the face of events that we do not control, we create alternative narratives that may suggest a reassertion of control. In this case, despite the resonance of images of destruction emanating from Alberta's bituminous sands, the narrative of technological progress suggests we are still in control. On the other hand, such alternative narratives may suggest ways of living in a world that we do not control. The chapters that follow, thus, might be read as debates between the upward counterfactuals of industry and government—which, while claiming to be simply factual, construct a future in which the trauma of the present leads to a better world—and the downward counterfactuals of the literary texts in which the sacrifices required by current bitumen extraction also sacrifice the future.

Chapter 1, "Lyric Oil," explores the potential for "lyric thinking" (Zwicky) to form the basis of a counternarrative to the hegemonic justification for bitumen extraction via readings of Rudy Wiebe's short story "Angel of the Tar Sands" and his play, co-written with Theatre Passe Muraille, *Far as the Eye Can See*. Both of these texts question human mastery over nature and the future, and they suggest that the unquestioned narrative of progress can be re-presented and questioned through a process of attentive listening.

Chapter 2, "Oil Sacrifices," uses Benedict Anderson's and Jean-Luc Nancy's theorizations of community to elucidate how various forms of sacrifice are justified in the name of bitumen extraction and to consider what the limits of those sacrifices should be. It contrasts the upward counterfactual of Sidney Ells's bitumen poetry with the downward counterfactual of Mari-Lou Rowley's "In the Tar Sands, Going Down" and Thomas King's "Alberta Oil Sands Land."

Chapter 3, "Impossible Choices," reads Marc Prescott's play *Fort Mac* and government and industry justifications for bitumen extraction alongside Harry Frankfurt's theory of "bullshit" and Bruno Latour's notion of "matters of concern" as a way of shifting the dialogue about bitumen to ecological questions. In the play, the audience is exposed to Kiki's trauma as she attempts to protect others from the forces of greed and domination that motivate many in Prescott's Fort McMurray. The play's conclusion opens the possibility of making Kiki's sacrifice meaningful through creating a different future, of taking the downward counterfactual of her death and transforming it into an upward counterfactual through its impact on the audience.

Chapter 4, "Irrational Oil," considers the role of free speech—*parrhesia*—and its limits in addressing the foundational challenges posed by bitumen extraction. It turns to literature as an alternative discourse that can present these challenges in a new way and how the irrational may be a necessary means of achieving this re-presentation (Cariou, "Tarhands" 17). Reading Warren Cariou's "An Athabasca Story" alongside the graphic book *Extraction! Comix Reportage,* this chapter considers the possibility for literary representations to increase the range of significations that unsacrificeable otherness can have within the discourse about bitumen. Instead of getting trapped in a binary where dead ducks (or environmental damage more broadly) either become a symbol of industry's commitment to "do better" (Katinas) or proof of its irredeemable environmental degradation, these texts attempt to perform a *parrhesiastic* (or truth-telling) function to shift their audience's relationship to bitumen but also use counterfactuals to serve this goal.

Chapter 5, "Pipeline Facts, Poetic Counterfacts," contrasts the figurative language deployed in two poetry collections opposing the Enbridge Northern Gateway Project with that used in the Joint Review Panel's report on the project, a Northern Gateway Pipeline advertisement, and journalistic accounts of bitumen extraction's environmental effects. It continues Chapter 4's engagement with the concept of *parrhesia* in regard to the importance of *ethos,* the character of the speaker, for rhetorical effectiveness and also considers the importance of the temporality of this effectiveness.

Chapter 6, "Oil Desires," reads "On the Wings of This Prayer" and "The Fleshing" from Richard Van Camp's *Godless But Loyal to Heaven* along with C.D. Evans and L.M. Shyba's satirical novel *5000 Dead Ducks: A Novel about Lust and Revolution in the Oilsands,* as returning to the questions asked in Rudy Wiebe's "The Angel of the Tar Sands." While Wiebe's story asks readers to choose the type of future their work will build, Van Camp imagines a future where unfettered bitumen extraction has already occurred before calling on readers to act now to prevent that future and the zombie race it will awaken. Evans and Shyba's novel imagines a parallel universe in which the names have changed but society's vampiric relationship with petroleum remains the same. The novel shows bitumen extraction having apocalyptic results and calls on readers to change their relationship to oil.

In the Conclusion, I consider again the power of industry and government narratives to represent bitumen extraction as the solution to the problems it creates. I look at the film *The Suncor Ankylosaur* as an attempt to counter, or neutralize, the counterfactual power of the literary counterfactuals explored in the preceding chapters. That is, the film works to reassert the power of industry, of human control over nature, to secure both the future and the past.

Unsustainable Oil attempts to contribute a new line of critique to the ongoing resistance to bitumen extraction. It draws on a wide range of texts from diverse genres—poems, plays, short stories, novels, as well as journalism, advertising, government and industry reports—in an attempt to explore the cultural terrain that bitumen has colonized. However, *Unsustainable Oil*, for the most part, does not address film as a medium. There are two main reasons for this exclusion: first, film is not my area of expertise; second, there are many films about bitumen, and engaging with these films in any depth would require another book-length project.

In writing *Unsustainable Oil*, I have tried to make use of my privilege—a privilege resulting from being born in a place and time where oil revenues were plentiful—to recognize, honour, and repay (at least in some small way) "all my relations" to whom I owe infinite debts (King, *All* ix).

1

Lyric **Oil**

Re-presenting Fossil Fuels As a Cultural By-Product

BITUMEN IS CENTRAL to the current narratives of Alberta and Canada. In the struggle to represent what is happening in and around Fort McMurray, Alberta, the official dominant story requires acceptance of bitumen extraction and export as a good thing. On December 20, 2012, Alberta Finance Minister Doug Horner, facing a budget deficit, "hinted that tax increases could be on the horizon if Alberta oilsands producers can't get access to new international markets" (Kleiss A1). Royalties on resource revenues fund basic government services. When resource revenues decline, Alberta's booming economy starts to go bust. Since "Alberta oil sells [to the United States] at a growing discount," the government solution is to help producers "get their products to Asia, where they might get higher prices" (Kleiss A1). The narrative is straightforward: increase resource revenues or face cuts to services and/or increased taxes.

This narrative is linked, I argue, to a foundational North American faith in progress, through the mastery of nature, to create a just and equitable world. This is, no doubt, a persuasive narrative. However, it is based on the abstraction of concrete diversity into the realm of commodities that can be bought and sold, and, as such, depends on citizens accepting that anything may be sacrificed if the price is right. Despite the dominance of this narrative, alternatives persist. The challenge is to create space to listen to these voices; creating such a space is one of the hopes of this book. *Unsustainable Oil* considers some emerging counternarratives, narratives that may displace the droning monologic justification of the status quo through the idealization of "sustainable development," which has dominated in Alberta, and Canada, for many years. This chapter considers industry and government rhetoric, which seeks to speak for both humans and non-humans in justifying the analysis required to separate bituminous sands into their component parts, and, thereby determine what the future of bitumen will be; it juxtaposes that rhetoric with an exploration of Jan Zwicky's concept of "lyric" undertaken through a reading of two texts by Rudy Wiebe.

Re-presenting Bitumen to Listen to Alternatives

Rudy Wiebe's "The Angel of the Tar Sands" and *Far as the Eye Can See* counter government and industry rhetoric by re-presenting those dominant narratives so that what they exclude becomes apparent. Following Gayatri Spivak, I want to note the difference between "representation as 'speaking for,' as in politics, and representation as 're-presentation,' as in art or philosophy" (*Critique* 256). "Speaking for" tends towards appropriative ventriloquism, while "re-presentation" offers the possibility of new understanding and insight. While it is ultimately impossible to maintain this distinction, moments of re-presentation are necessary if there is to be any change in the relations between those who are speaking and those being spoken for. Literature can and has re-presented the necessity of ecological coherence and the threats posed to it by industrial capitalism; we have a responsibility to listen to these alternatives and live our lives differently after we've heard these stories (King, *The Truth* 29).

Literature can offer a practice in vulnerable listening, attending to otherness, the marginalized, the unsayable—as in Wiebe's texts. In his short story "The Angel of the Tar Sands," we meet the superintendent of a bitumen project, whose certainty about the need for particular actions to keep the separation facility operating at capacity are momentarily called into question when an angel is unearthed in the bitumen deposit. In his play *Far as the Eye Can See,* the defeat of a proposed coal mine due to local opposition and the last-minute intervention of Peter Lougheed—Progressive Conservative premier of Alberta at the time of the play's creation—is suffused with irony over the political pragmatism of Lougheed's decision. In the story, the superintendent's silence when confronted with the angel requires the reader to attend to the angel's alterity. In the play, the conclusion's irony requires the reader or audience member to attend to the persistent threat posed by technocrats and the world they seek to create.

Literature, then, is one place to look for alternative narratives that reveal what is left out of progressive and teleological myths, alternatives like that suggested by Wiebe's angel. Even if literature does not tell us what we should do, it can suggest that something is missing in what we are currently doing. Speaking at the University of Alberta in October 2007, poet and philosopher Jan Zwicky argued that "technocracy prevents us from living authentically if we follow its agenda." She suggests "lyric" thinking as a way of accessing what is left out of purely rational-analytical thought, what is left out of thought focused on extracting oil from sand by any means necessary. She explains elsewhere that her

> use of "lyric" quite deliberately sets aside surface historical associations with Romantic poetry in order to pursue what might be called its deep epistemological structure. The word has been used to characterize things as disparate as Vermeer's paintings and Schubert's use of diatonic tonality, and what is common to them all...is thought whose *eros* is coherence. (*Wisdom* n.p.)

Zwicky explains that she does not use "lyric" "in a sense that emphasizes the role of the individual ego: the 'outpouring of subjective emotion'...That sense is corrupt and is based on a subversion of the desire which fundamentally underlies lyric expression—relinquishment of the individual ego rather than celebration of it" (*Lyric* 126). Therefore, one of the things lyric thinking can help us recognize is that ecological value, the value of functioning wholes, cannot be reduced to either use or exchange value, is always beyond the humanly convenient. Its eros, its desire, is to recognize the interconnectedness of human and non-human nature, the relatedness of parts, and the coherence of the whole. Human concerns—culture, economy, society—exist within a larger order, one in which human needs are not primary. "The Angel of the Tar Sands" and *Far as the Eye Can See* suggest what remains outside anthropocentric narratives and point to the coherence of which humans are only a part. Before considering what the story points beyond, though, we need to establish the field in which it is intervening.

Overcoming the Burdens of Nature

The massive and various displacements required by the bitumen extraction projects surrounding Fort McMurray—of land, labour, and prior communities—continue to be justified by a political representation of a "common good" of prosperity, equality, and independence that is on the way and can be achieved through work. For example, Leslie Shiell and Colin Busby, writing for the Canadian economic public policy think tank the C.D. Howe Institute, argue that "all Albertans, including those yet to be born, are entitled to an equal share of resource-based spending" (3). This guarantee of continued "unprecedented prosperity" (1), though, depends upon sacrificing the finite ecological diversity on which all life, including the human, depends. The process of extracting oil from sand involves removing "overburden" to access bitumen;[1] burning natural gas to create the necessary heat to separate oil from sand; drawing water from the Athabasca River for use in various extraction processes, some of which actually sterilize the land (Shukin, "Animal" 136; L. Pratt, *Tar Sands* 16); constructing tailings ponds for the now toxic water polluted with the by-products of upgrading the oil. David Finch elaborates:

> To get one barrel of oil, 2 tonnes of tarry sand have to be *dug out* of the ground...Some bitumen goes to the plant through a pipe line, after being *broken up* and mixed with hot water. When the oily sand gets to the plant, hot water helps *release* the oil from the sand as the whole mixture gets *shaken* and *stirred*. Bitumen *rises* to the top and *goes off* for further processing, and more oil gets *squeezed out* of the slurry. Even then, it still has to go through an upgrader, which makes it lighter using diluents or lighter oil, before it can go into a refinery. (111, emphasis added)

The verbs in this passage highlight that this method is a form of analysis, defined as "separation of a whole into its component parts" (Zwicky, *Lyric* 5). Humans are very good at analysis, as the "success" of the bitumen industry shows, but once the rich complex of something whole is taken apart, dis-integrated, it cannot be put back together. The factual "success" of analysis is used to support the counterfactual claim that reclamation is possible. At the same time, bureaucratic doublespeak about reclamation implicitly recognizes this inability when it states that "companies are required to replace what was there before industrial activity with a landscape of 'equivalent capability.' This means the land should be able to support similar, but not necessarily identical, uses after reclamation is complete" (Brooymans, "Oilsands' Newest Project" A1). Such a focus on "uses" ignores the ecological point of integration: value lies in functioning wholes as well as in the production of useful or exchangeable parts. Regardless of this narrow and troubling definition of reclamation, it is worth noting that, as of June 2015, only 104 hectares have received certification as reclaimed, and these 104 hectares were never mined (whereas, over 89,500 hectares in total in the area *had been disturbed*). Rather, the 104 hectares were used to store the overburden taken from land being mined.[2]

The crucial point, for my purposes, is that once the dis-integration of analysis has occurred, there is no system available for reintegration. Although the Alberta government requires bitumen companies to pay a security deposit against the cost of reclamation, the "companies calculate the cost of their own security deposits," and there is no opportunity for public input in this process (Grant, Dyer, and Woynillowicz 46). Alberta

School of Business Professor Andrew Leach writes about this issue: "The Alberta government's revision of the Mine Financial Security Program (MFSP) continues down a wrong-headed path where the province is willing to take on environmental risk to enable oil sands development" ("A False Hope"). Further,

> Under the program, operators would initially provide a "base amount of security for each project," which would be sufficient only to secure but not reclaim the site. Full financial security would not be required until six years prior to the end of operation of a facility. This means that, for the first 50–60 years of operation, the revised program would still allow mine sites to impose unfunded environmental liabilities on the province. (Leach, "A False Hope")

It is important to note, as Pembina researchers Jennifer Grant, Simon Dyer, and Dan Woynillowicz do, that "the fact that virtually no reclaimed sites...have been certified suggests not only that the true costs have *not* been identified, but that they *cannot* be identified" (46, emphasis original). Their subsequent recommendation, that "liability bonds [should be] conservative to ensure that Canadians are protected" (46), thus rings hollow. If the true costs of reclamation cannot be identified, then the only way Canadians and others, not to mention non-human others, can be protected is by not dis-integrating the sites to begin with. If humans continue to pursue fiscal independence through resource extraction, we incur an incalculable debt as a result.[3] The consequences are unaccountable.

The dominant progressive narrative puts nature at humans' disposal. This narrative is apparent in histories of the Fort McMurray region. These tend to start with accounts of a man named Wa Pa Su (or Wa Pa Sun or Wapisiw[4]) bringing "a sample of oil laden sand" to Hudson's Bay Company explorer Henry Kelsey at York Factory in 1719 (Humberman 4). This moment of first recorded contact with the indigenous inhabitants of the area does two things. It provides an origin and teleology for the narrative of the region, one that will reach its climax in our own time when the bitumen can be displaced from the sand. It also implies continuity with indigenous practices: Natives burned the sands and used them to seal their canoes; contemporary uses are simply logical and natural progressions of indigenous ones, operating on a larger and more efficient scale. Robert Bott, in "Canada's Oil Sands," expresses it this way: "Aboriginal people were already using oilsands bitumen when the first European explorers arrived in the 18th century" (11). In Wiebe's play *Far as the Eye Can See* (1977), William Aberhart and Crowfoot—who, along with Princess Louise, make up the "Regal Dead"—discuss Calgary Power's plan to expropriate farmland and build a coal mine. Wiebe has stated that the current debate surrounding the oil sands is "exactly the same issue" as the fight over the Dodds-Round Hill power plant represented in *Far as the Eye Can See* (Babiak):

> CROWFOOT: A hundred years ago this year we gave this earth away and now the white man has at last decided he will chew it up and spit it out like a beast eating itself.
>
> ABERHART: No no no, chief. You never knew about all the food this land could grow, the grain, the cattle, the beautiful gardens, the potatoes and flowers.

Crowfoot responds with a wonderfully understated line—"We had flowers"—which points toward one form of otherness that is threatened by the colonial economy, when monocultural agribusiness replaced the bison economy. Aberhart, however, does not pause to consider the implications of Crowfoot's statement. He plunges on with this point:

> ABERHART: All right, flowers, but the power to grow all these other things was already there in the soil, waiting to be released. And we whites found it there, released it! Fed millions, all over the world. Now this other enormous power of coal, waiting to be released, placed there by God, is waiting to be released to lift man to his next stage of progress. These Albertans, in 1977, will develop the treasures *under* the earth, as my generation reaped them from the earth, as your people gathered them upon the earth. It is all there, the marvelous gifts of God, for us to use, for us—(68, emphasis original)

This progressive logic continues to operate in narratives like Bott's, which connect current and previous actions through a focus on *use*. However, while Bott's and Aberhart's version of history may suggest some recognition of "the dependence of oil sands capital (like fur trade capital before it) upon native informants/knowledges," nevertheless, "this debt is relegated to the distant past and ultimately reversed as corporations such as Syncrude depict themselves as benefactors bringing economic development to Aboriginal communities" (Shukin, "Animal" 144). Thus, while maintaining continuity and establishing indigenous validity for using the oil sands, "First Peoples' knowledge of the oil sands is relegated to the past, insinuating that true knowledge of the oil sands begins at the moment when it ceases as use-value and is measured in terms of exchange-value" (Shukin, "Animal" 145). This abstraction is fundamental for the system of modernity, which "*consists in turning the productive origin inside and representing it as simply another factor within the system*" (Angus, *Border* 188, emphasis original). This turn radically reconfigures humans' relation to the non-human world: instead of owing our existence to it, it is an object to be bought and sold by us. Hence the irony of the conclusion of Wiebe's play: the crisis is resolved, or, at least, deferred,[5] by Peter Lougheed descending in a coal bucket as *deus ex machina* to protect agricultural land over electricity generation. Rather than Lougheed saving the day, though, the stage directions suggest that "nothing human has been resolved by this [Lougheed's] mechanical statement" (124). His statement is a rhetorical performance aimed at the "voters" (124) of Alberta, and while Lougheed declares his "government is committed to provide every citizen with adequate electricity...this project would disturb too much prime agricultural land" (124). The decision is a human-centred one, defending one type of human land use over another. The play ends with these lines:

> CROWFOOT: A hundred years ago the white man took this country, and now they know the black burning stone is under the earth; and they will not be able to leave it alone. White men can never leave, anything, the way it was.
>
> LOUGHEED: We have listened to you, and we have understood you. Thank you. (125)

The irony is multiple here, pointing to the dominant group's belief in its understanding of the Native worldview, a paternalistic dismissal of that worldview, and a suggestion that such a dismissal will create precisely the situation Crowfoot foresees: us eating ourselves. While Lougheed fails to really listen or understand, the play offers the opportunity for the audience members to listen and to recognize, in their own views of the world-as-resource, the hubris of Lougheed's response. Stopping a coal-fired generator development to protect agriculture leaves one form of commodification of the land in place, but it does not amount to leaving something the way it was. Lougheed does not see an alternative to a version of private property and commodity markets.

The cover letter from the Alberta Royalty Review Panel's report *Our Fair Share* shows the logic of this commodification at work. It structures the problem of royalty rates in terms of owing, but, instead of considering what humans owe the non-human world due to our objectification of it, it considers what is properly *owed* to Albertans, the *owners* of resources. Chairman William M. Hunter writes, "Alberta's natural resources belong to Albertans...A royalty and tax system for energy resources therefore must justify every dollar that does not go to the owners" (5). This language always already introduces a provisional understanding of the concept of *owing*, one inextricable from *owning*. The possibility of "responsible" or "sustainable" development of Alberta's oil and gas requires a more radical understanding of owing, one that is not contractual or provisional.[6] Re-presenting such an understanding so that people will listen is increasingly difficult, but it is not impossible.[7]

"Beyond All Bargains"

In the November 2007 issue of *Alberta Views*, a number of pieces focus on the effects of development in Northern Alberta and point to what is sacrificed by the provisional views justifying that development. In "The Standard Is Zero," Dr. Rosalie Bertell critiques the Cantox Health Sciences International assessment of the North West Upgrading Project:

> Regulatory guidelines are risk-versus-benefit trade-offs, with little or no protection for the public against mistakes, miscalculations or even "acceptable" cancers and deaths. Courts and most people see regulations as if they were protective of human and environmental health...[but] Regulations are frequently based, as is the case here, on "acceptable" deaths to humans and unknown effects on the environment. (19)

While such trade-offs are standard practice in our modern, pragmatic political system, where the only protection against negative consequences is to make sure we are adequately compensated by receiving "our fair share" in taxes and royalties, we can hear in Bertell's words an echo of an alternate conception of responsibility. An understanding of this echo can be found in the work of Canadian philosopher George Grant. He has, arguably, thought through the consequences of a fully realized technological society in the Canadian context more completely than anyone else. In his last collection of essays, *Technology and Justice,* he writes that

> the limitations put upon creating by the claims of others, whether nationally or internationally, are understood as contractual: that is, provisional. This exclusion of non-provisory owing from our interpretation of desire means that what is summoned up by the word "should" is no longer what was summoned up among our ancestors. What moderns hear always includes an "if": it is never "beyond all bargains and without an alternative." (30)

This "should," which is beyond any "if," is hinted at in Bertell's words, but such a conception of responsibility is continually silenced in public discourse dominated by the rhetoric of resources, a rhetoric that understands our relationship to the world as one between human freedom and natural raw material.

The consequences of this view are that human and non-human nature can be, and is, sacrificed within the logic of the market. Rather than maintaining that some actions should not be taken under any circumstances, anything may be done if the price is right. Grant summarizes Nietzsche:

> Politics is the technology of making the human race greater than it has yet been. In that artistic accomplishment, those of our fellows who stand in the way...can be exterminated or simply enslaved. There is nothing intrinsic in all others that puts any limit on what we may do to them... Human beings are so unequal that to some of them no due is owed. (*Technology and Justice* 94)

While extermination and enslavement might be stronger language than contemporary bureaucrats, technocrats, and capitalists are willing to use, the exclusions and exploitations that even so-called sustainable resource extraction requires amount to the same thing for some. Jeremy Klaszus writes of the health problems of the population of Fort Chipewyan arising from pollution of the Athabasca River: "Premier Ed Stelmach famously said there's no such thing as touching the brake on oil sands development, and the environmental concerns of people in Fort Chip aren't significant enough to slow anything down" (32).[8] Since Klaszus published that article in November 2007, the Chipewyan Prairie First Nation, the Woodland Cree First Nation (signatories to Treaty 8), the Beaver Lake Cree First Nation (signatories to Treaty 6), and the Lubicon Cree (who have never signed a treaty) (Lillebuen; Stolte) have filed claims arguing that development jeopardizes treaty guarantees "to continue hunting, trapping, and fishing for their livelihoods" (Stolte).[9] On October 5, 2012, the Chipewyan Prairie First Nation launched a suit against the provincial government and Shell Canada for failing to consult adequately over the proposed expansion of Shell's Jackpine mine (Ball). As early as June 12, 2009, the *Edmonton Journal* reported, "The amount [of money] Alberta owes First Nations affected by oilsands development could easily outstrip all the royalties the province has earned off the resource" (Stolte B6), if the lawsuits are successful. These lawsuits may turn bitumen projects from money-making into money-losing ventures, and, thereby, slow or stop the development. More recently, coinciding with the Idle No More movement in 2013, the Mikisew Cree First Nation and the Frog Lake First Nation launched a judicial review of the federal government's omnibus budget bills C-38 and C-45 because of lack of consultation on the legislation's changes to the

Fisheries Act, the Environmental Protection Act, and the Navigable Waters Protection Act (Kennedy, Ignasiak, and Duncanson). In April 2013, the Fort McKay First Nation filed an objection to a steam-assisted gravity drainage (SAG-D) project proposed by Dover Operating Corp. This is "the first time in 30 years [that] Fort McKay First Nation has filed a formal objection against an oilsands project on its traditional territory" (Wohlberg). As of June 2015, none of these lawsuits has been successful and several have been dismissed, but, in April 2013, the Alberta Court of Appeal dismissed a second attempt by the federal and provincial governments to have the Beaver Lake Cree's case thrown out. The court upheld the grounds for the case, which argues that the cumulative effects of bitumen projects in their territory are "having a negative impact on their rights under Treaty 6 to hunt and fish" (S. Pratt, "Appeal" A6). The case can now proceed to trial.

Increased health problems for residents are not the only social consequence of industrial growth in the oil sands. Cheryl Mahaffy describes the lives of prostitutes in Alberta's boomtowns: "Often driven by poverty, addictions, or both, and unable to find affordable housing...most are homeless or barely housed. They lack access to showers, let alone health care" (38).[10] The standard of suffering is not zero. What royalty rate would be fair compensation for these consequences? Such a question can only be asked when certain assumptions are taken as self-evident. If we start from the premise that "there is some connection between what we think other species to be (let alone our own) and how we treat them" (Grant, *Technology and Justice* 65), this connection becomes evident when we examine the language of sustainable resource development: other species, and members of our own, are objects to be used, and, if anything is due to them, their extermination or enslavement can be compensated for with a share of the profits.

Responding, in the *National Post*, to the public outcry over the April 29, 2008 deaths of 1,600 ducks on Syncrude's Aurora tailings pond, Ezra Levant trivializes citizens' concerns and re-inscribes a utilitarian view of nature as "common sense": "even the Prime Minister called the deaths a 'terrible tragedy.' I agree. I can't believe that 500 ducks were wasted that way. Imagine all the lost foie gras." Never mind that foie gras is made from goose liver, Levant's sarcastic point seems to be that there is no consequence of development that cannot be compensated for, and the only tragedy is that the ducks' lives were lost without being commodified.[11] While public discourse allows for debate over what the human relationship to oil in Northern Alberta should be, the primacy of that relationship is rarely questioned: the debate is organized by terms such as "sustainability," "responsible development," and "stewardship." Humans will manage the resource in our best interests; what these interests are and what the best management looks like are the debatable topics. However, these debates remain caught within a liberal, rational, technological, humanist narrative of progress founded on a premise of autonomy over a future in which the productivity of nature will be guaranteed through increasingly efficient systems of management. Attempts to tell stories other than that of human autonomy are rare and marginal.

Internalization of productive natural origins is expressed in the economic term *fungibility*, which Levant describes without naming: "If oil-thirsty America—and China and India—stop using oilsands oil because it's duck-insensitive, they're just going to buy their oil from somewhere else" ("Shed No Tears"). That is, any barrel of oil is economically equivalent to any other. However, while this may be true on the global market, other factors exceed the market's accounting. Again, Levant notes some of these as justification for bitumen development:

> Pretty much everywhere oil comes from is worse than Canada—even with our new reputation as duck murderers...most oil producing nations are dictatorships in the Middle East, or quasi-dictatorships like Nigeria, Russia and Venezuela. Political accountability doesn't exist there, and property rights are weak. The idea of political parties or even a free press championing the environment is a fantasy.

In distinguishing Canadian oil from "global" oil, he wants to have it both ways. Oil sands production is *indistinguishable* from global oil once it enters the marketplace, but it is also *better* than other oil because it does not depend on dictatorship, misogyny, genocide, etc. The fact that it does depend on these things, at least to some extent, is not considered: Mahaffey's analysis of prostitution in the area indicates misogyny;[12] the widespread indifference to the concerns of First Nations suggests a continuation of a colonial genocidal policy.[13] Levant's conclusion is simple: "It's true, the odd duck gets hurt—as with any other massive industrial project. But to portray the oilsands as unethical is positively false." And he encapsulates progressive modernist logic: the odd duck, or person, or ecosystem may get hurt, or killed, or become extinct, but that is not unethical *if* property rights are protected, and that means honouring contracts made with oil companies and, at best, getting "our fair share" as owners of the "resource." We *should* be concerned about the ducks, as Levant facetiously suggests he is, but only *if* it is in "our" best interest. That interest will be viewed through the political lens of "progress," and this view maintains that "environmental issues are largely local or regional in geographic scope while the economic benefits are provincial if not national [or, indeed, global]" (Babiuk 13). "The Angel of the Tar Sands" points to a context beyond the national or even the global that needs to be considered. The "unscientific shape" of the angel disrupts the idea that environmental issues are simply local (189).

Within the progressive view, the productivity of nature, its fecundity—on which systems of exchange are founded when land, for instance, becomes a commodity—is partially captured within systems as a profitable yield (Angus, *Border* 197). However, the process of capturing these yields does not, and cannot, account for all of the productivity; an uncommodifiable excess, something unaccountable, like the angel, remains. Ian Angus says, "Human domestication of the excess that defines the wild shapes the wild order into a humanly convenient one, reduces diversity, and takes the excess for itself" (*Border* 197). But, while "excess keeps the system going," it "finds no recognition within it, only the belated and diluted recognition of a yield" (Angus, *Border* 197). The natural productivity is not recognized as an origin, simply as one commodity among many; the failure of the characters to account for the angel, to fit it into the system of commodity exchange, points to a need to consider the world in terms that are not merely humanly convenient. The idea of reclamation in the Alberta oil sands explicitly perpetuates this bias towards human concerns, as we will see below. While many environmental lobbyists, think tanks, and policy analysts are publishing important critiques of industry and governments' ecological conduct, their alternatives tend to be politically pragmatic calls for improvements to the system rather than recognition of what inevitably escapes systematization. The sacrifice of this excess needs to be recognized if we are to determine what we *should* do, beyond all the bargains that have been struck and the profits to be made. Lyric thought offers one mode of attending to this sacrifice.

A Lyric Alternative to Objectifying Reason

Zwicky contrasts the dis-integration of analysis with lyric: "The aims of analysis are explicitly disintegrative" (*Lyric* 6), whereas "lyric is an attempt to comprehend the whole in a single gesture" (134). The primacy of analysis over lyric results in "the ability to see some thing as an object, external to oneself," which "is necessary before one can cease to care about that thing sufficiently to regard it as nothing more than an object one can use" (234). Thus, the ecological diversity of the Athabasca River watershed is seen as useful because of the marketable objects it is capable of producing. Since wild ducks, for example, are not competitive with oil in a global commodities market, and, as Levant has it, they are "a species that is as far from extinction as, say, the rat," their sacrifice is acceptable.[14] For Grant,

> Object means literally some thing that we have thrown over against ourselves. *Jacio* I throw, *ob* over against; therefore, "the thrown against"... Reason as project (that is, reason as thrown forth) is the summonsing of something before us and the putting of questions to it, so that it is forced to give its reasons for being the way it is as an object.
> (*Technology and Justice* 36)

This, he says, results in a paradigm in which "we have knowledge when we represent anything to ourselves as object, and question it, so that it will give us its reasons. That summonsing and questioning requires well-defined procedures. These procedures are what we call in English 'experimental research'" (*Technology and Justice* 36). A great deal of "experimental research" has been, and is being, thrown against bitumen to force it to give its reasons for being the way it is and to enable its analysis and separation.

Objectifying reason, however, ignores the degree to which we are integrated with what we objectify. Ecologist Stan Rowe argues, "People are made from clean air, clean water, and clean humified soil—all those things that exist in Earth's green surface film. To these their healthy bodies are adapted" (117). However, he continues, "*Homo sapiens* ignore this fundamental knowledge, secure in the belief that human errors can be...cured by science and technology" (117).[15] As a result, humanity "makes a virtue of mining and releasing into air, water and soil the many poisonous materials that used to be safely sequestered underground: heavy metals, radioactive minerals, long-chain hydrocarbons such as coal and oil" (117). Former Syncrude president Jim Carter, in contrast to Rowe's critique of techno-utopianism, asserts, "The solution [to ecological disturbance] is in technology" (qtd. in Marsden 166). Although recent pro-industry representations have worked to create a more complicated framing of the challenges and virtues of bitumen extraction, this narrative of techno-utopianism continues. Gismondi and Davidson write, "more contemporary counter frames are more nuanced, framing the tar sands as not exactly clean but less dirty than alternatives, as a problematic but necessary development in an energy-hungry world, and for some, a means of transition to a new energy age" (203). However, despite an increasing recognition of the negative ecological consequences of bitumen extraction, technological optimism continues to reign supreme. For example, in response to a report by Environment Canada scientist Derek Muir and Queen's University biologist John P. Smol published in the *Proceedings of the National Academy of Sciences*, which shows that "production in the oilsands is polluting air and water at greater rates than previously thought" (Klinkenberg, "Rethinking" E1), former Syncrude CEO Eric Newell claims, "Technology got us where we are today, and it will take us where we need to be in the future" (Klinkenberg, "Rethinking" E1).[16] That is, the cause of the problem is also the solution; further objectification and dis-integration will create "sustainability," despite the fact that, as William Marsden states, "Nobody at Syncrude claims that they can restore [the land] to what it used to be. They can't replace the ecosystems, the vast networks of peat bogs, fens, rivers and wetlands. Those are gone, probably forever, along with the unique flora and fauna that once inhabited this region" (168). Those ecosystems, part of the life-sustaining "green surface film" redefined as burdensome overburden, are sacrificed for the bitumen beneath. However, the acknowledgement of this fact by Syncrude takes a particular form: they assert that reclamation makes the land "more productive than it was when we got here" (Marsden 170). Marsden quotes Peter Duggan, a production advisor for Syncrude, as saying the company is "doing nature a favour" by "cleaning out this dirty oil out of the oil sands" (170).[17]

Similarly, the Government of Alberta's wetland policy recognizes the impossibility of recreating the systems displaced by bitumen extraction. On September 10, 2013, a new policy was announced to protect wetlands in the province. However, this policy "moves away from the no-net loss principle" and creates "a hierarchy of wetlands with a policy geared to offer more protection to 'high value' wetlands" (S. Pratt, "Policy" A4). Further, while "Oilsands developers and other industry would be asked to avoid or minimize negative impacts on wetland...if that's not possible, they could be asked to replace a wetland near the original area or pay into a fund" (S. Pratt, "Policy" A4). This policy has been eight years in the making and it is expected to take two more before a yet-to-be established committee evaluates the relative value of wetlands. An expert from Alberta Environment, Thorsten Hebben suggests, "In the long run it's possible this policy would allow wetlands to migrate from the northeast [i.e., the bituminous sands region] to areas deprived of wetlands such as ranching areas in the south and agricultural land like the Peace River farming area" (S. Pratt, "Policy" A4). This seems to suggest that there are regions with more wetlands than they need and human management can somehow destroy wetlands in one area and create them somewhere else without disrupting the ecological relations in either: the framing of this attempt to control nature in terms of "migration" strikes me as particularly manipulative.

The human capacity to create moments of resonance and integration even out of waste and detritus should not be underestimated, as Patricia Yaeger argues in the March 2008 edition of PMLA. However, this does not mean that we should "drop the concept of nature" (Morton qtd. in Yaeger 324). There is a necessity, in the postmodern condition, to reveal the aesthetic beauty possible within even the most "manufactured landscapes"—to borrow the title phrase of Jennifer Baichwal's 2006 film based on Edward Burtynsky's photographs—and the persistence of wilderness, as that which precedes, exceeds, and succeeds human rational-analytic projects, among our garbage. Nevertheless, though waste and debris may well be able to do the "subject-making" work once accomplished by a now marginalized and always already contaminated "nature" (Yaeger, "The Death of Nature" 325), this perspective only recognizes nature's value for doing a kind of traditionally human-oriented integrative work associated with a kind of Romantic "lyric."

Zwicky wants to set those associations aside in favour of a higher order context. At the same time, while wilderness will persist, regardless of humans' destructive activities, something of more than contingent value will also be lost. Dropping the concept of nature opens the door further to a view of the world in which nothing can be said "as to why we should be glad that there are polar bears or parrots or why they are worth working to preserve" (Grant, *Technology and Justice* 64). Dropping "nature" furthers a purely contingent worldview, which enables a belief that the boreal ecosystem of Northern Alberta is actually *improved* by the strip-mining of the boreal forest to extract the bituminous sands beneath.

Slowing the pace of development, though, is an answer of *degree* to a problem of *kind* as it advocates "sustainability" or "responsible development," which Okanagan writer Jeanette Armstrong described in a 2007 lecture as "another way of (ab)using such that enough remains so that we can continue to (ab)use." It substitutes one form of analysis for another. Armstrong suggests an alternative of "unqualified regeneration," which requires recognition of human dependence on and interconnection with the ecological whole. Such recognition is incompatible with the belief in autonomy asserted in Syncrude's desire to "create [its] own future," a phrase that appeared in the company's sustainability reports from 2003 to 2006 (*Focused: 2006 Sustainability Report* 11). The power to create one's own future suggests independence from the web of biodiversity that supports human life, which resonates, for me, with the words of a senior Bush advisor who in 2002 said, "We're an empire now, and when we act, we create our own reality" (qtd. in Coleman 40).[18] Both statements indicate a pragmatic understanding of truth: "Our ideas are true," states George

Grant in his critical definition of pragmatism, "when and insofar as they are effective in action" (*Philosophy* 83).

The human power of creativity can work for or against such discourses of mastery. However, "ethical self-mastery, political mastery over unruly and aberrant Others, and epistemic mastery over the 'external' world pose as the still-attainable goals of the Enlightenment legacy" (Code 19). The narrative of this legacy remains persuasive, and both sides in the debate around bitumen have tried to claim its rhetoric as, for example, when Christopher Hatch and Matt Price, authors of the Environmental Defence report *Canada's Toxic Tar Sands: The Most Destructive Project on Earth*, write, "Technologies are available to curb the damage" (2), or when Marcel Coutu and Scott Sullivan write in Syncrude's *Are the Oil Sands Being Responsibly Developed?: 2010/11 Sustainability Report*, "Through collaborations with academia, industry and other research-oriented organizations, we are making significant progress but recognize there is more we can do as our capabilities and technologies expand" (2).

Sustainability is a key term for both sides and has become, in an Orwellian way, meaningless. However, if "more responsible knowings than the reductionism endemic in the positivist post-Enlightenment legacy" are to arise, then what is excluded by this reductionism needs to be recognized through what Spivak calls re-presentation (Code 9). If we can begin from a recognition of what is excluded and sacrificed by mining's analytic dis-integration, then we can have a different kind of public debate over bitumen's future. A debate that puts regeneration (as restoration of natural productivity not oriented to human ends) rather than reclamation or "sustainability" (as manufacturing a new kind of profitable landscape to be exploited) at its centre may avoid positivist reductionism. Rudy Wiebe's "The Angel of the Tar Sands" points us toward such a recognition by calling attention to what is left unspoken by the dominant narrative of technological progress. This story also highlights what remains unspeakable for many of its critics: our own vulnerable, limited, dependency.

Given the literary challenge of coming to terms with the geographic vastness of Alberta, Wiebe has famously stated,

> to touch this land with words requires an architectural structure; to break into the space of the reader's mind with the space of this western landscape and the people in it you must build a structure of fiction like an engineer builds a bridge or skyscraper over and into space. A poem, a lyric, will not do. You must lay great black steel lines of fiction, break up that space with huge design, and like the fiction of the Russian steppes, build giant artifact. ("Passages by Land" 259)

This might be, at least in the most straightforward reading, an accurate description of what some of Wiebe's earlier fiction does, with its tendency towards monologic didacticism. However, both his short story "The Angel of the Tar Sands" and his play *Far as the Eye Can See* do something else. As Jonathan Kertzer puts it, the story shows "how the matter-of-fact must yield to the marvelous" (xxiv), while the play's "ironic moments" (Whaley 46) arise from some characters' failure to understand anything beyond the matter of fact. The ironies in Aberhart's statement about the right to use the land however humans see fit and in Lougheed's claim to have understood Crowfoot's warning about leaving things alone, quoted earlier, suggest their inability to accept dependence on what is beyond the matter of fact. Zwicky describes the process of yielding to what is beyond the matter of fact in writing when she says, "lyric achieves integration only to the extent that words are bent to the shape of wordlessness" (*Lyric* 256). The building with words, the huge design, will not succeed in enclosing meaning in the structure of words, but can suggest integration by yielding to silence; a writer's design can break into the reader's mind by building around and toward the landscape. Thus, human creativity can work to touch the land but not enclose it, to build a structure of fiction but not in the service of domination. Such building is *lyric* in Zwicky's sense regardless of whether it occurs in prose or verse, music or architecture, paint or steel.

Still, the economic logic of development persuasively embodies the promise of prosperity and an end to human suffering and hunger. Robert Kroetsch, in *Alberta*, his history of the province, states the view succinctly: "Ultimately, I am told, petroleum products from places like the Athabasca tar sands might not only run our cars and surface our highways, but *feed us* as well" (274).[19] In "The Angel of the Tar Sands," the lyric moment occurs after the three characters in the story—Tak, Bertha, and the unnamed superintendent—encounter an angel. It is unearthed by Tak, "the day operator on Number Two Bucket" (188); Bertha recognizes the "shape" (188) as an angel; and the superintendent, when faced with something so "unscientific" (189), loses "all words" (190). This loss suggests a failure of reason, an inability to apprehend something beyond the rational understanding of bitumen mining as oriented toward human ends: fueling the economy, feeding human needs. For Zwicky, "Lyric is a direct response to the fact that the particular capacity for language-use possessed by our species cuts us off from the world in a way, or to a degree, that is painful" (*Lyric* 246). Further, "We experience *the burden of our capacity for language* as loss—though we rarely recognize that *this* is the burden, that what we have lost is silence" (*Lyric* 246, emphasis original). Recognizing this burden as loss is a precondition for lyric, and "Lyric art is the fullest expression of the hunger for wordlessness" (*Lyric* 246). This hunger, I would suggest, is acknowledged by Bertha at the end of Wiebe's story in her act of quitting her job: she is not willing to go on sacrificing whatever perfection she has apprehended in her encounter with the angel. Further, her assurance to Tak, "Next time you'll recognize it...And then it'll talk Japanese" (191), opens the conclusion out to the reader by suggesting that, like Tak, we should be ready for the appearance of the divine in our world, and, if we are, then we will be able to listen to it as it will speak our language. It also offers us the possibility to choose to change our relationship to bitumen. With Bertha, we can walk away rather than continue to sacrifice the perfection that Bertha experiences and which the story can intimate for the reader.

Bertha is the only character who can understand what the angel says, as it speaks "Hutterite German," but she is unable to translate that speech into English or even repeat it as she has "forgotten" how (190). Although it is not silence, as such, that she has lost here, she is rendered silent by the angel's speech: "She was listening with an overwhelming intensity; there was nothing in this world but to hear" (190). She is unable to speak, is connected to something beyond language. The angel's words are unrecorded in the story—we don't get the Hutterite German, don't know what the angel said—and so, even though we are told the angel spoke, its words are an absence, a silence in the text. We, too, are connected to something beyond language through the language of Wiebe's story. For Angus, silence "is the discovery of a moment beyond language, though the beyond of language cannot be discovered outside language, but is that within language that allows its outside to appear" (*(Dis)figurations* 249). The superintendent, though unable to understand the angel, at least glimpses this too as he replies to Bertha's "I quit...Right this minute" with "Of course, I understand" (191). However, he struggles against the silence imposed by the angel, attempting to speak and "pleading for a voice" (190), and his recognition is quickly covered over when he instructs Tak to literally cover the evidence of the angel's presence by "run[ning] [his] bucket through" the sand where the angel had been (191). The production of oil is of tantamount importance so that any other story is too costly to listen to: the superintendent initially fears Tak has found "*another* buffalo skeleton" and the "bifocalled professors" that it would bring (188). He will hide evidence of the land's alternative stories in order to maintain productivity. Tak acquiesces to this request, becoming complicit when he says, "You're the boss": he is just following orders (191). Bertha points out that "It doesn't matter how fast" they erase the evidence because "It was there, we saw it" (191). A parallel might be drawn here to the public response to the duck deaths on Syncrude's tailings pond (see Figure 2). People were presented with images of death and suffering on a relatable scale clearly and directly tied to bitumen extraction and associated industrial activity. However, the initial emotional response passed through a combination of rationalizing, as with Levant (there are millions of ducks and humans need oil), and apologies, with Syncrude's then-president and CEO Tom Katinas's "promise to do better."[20]

In Wiebe's story, the return to "business as usual" does not quite occur immediately. First, the superintendent "had a vision" (191):

> He saw like an opened book the immense curves of the Athabasca River swinging through wilderness down from the glacial pinnacles of the Rocky Mountains and across Alberta and joined by the Berland and the McLeod and the Pembina and the Pelican and the Christina and the Clearwater and the Firebag rivers, and all the surface of the earth was gone, the Tertiary and the Lower Cretaceous layers of strata had been ripped away

FIGURE 2: *Syncrude tailings pond.* [David Dodge, the Pembina Institute]

FIGURE 3: *Suncor mine.* [David Dodge, the Pembina Institute]

> and the thousands of square miles of black bituminous sand were exposed, laid open, slanting down into the molten centre of the earth. (191) (see Figure 3)

In this one moment, which encompasses the entire Athabasca River watershed in a single sentence, the superintendent feels the meaning of the words of Psalms 51—"*O miserere, miserere*" (191)—in which David seeks God's mercy and forgiveness for his sins. This moment relies on a particular Judaeo-Christian conception of the world as David comes face to face with the particular devastation he has caused, and repents in humility. The question of complicity must be raised here. For Grant, "what has made Western culture so dynamic is its impregnation with the Judaeo-Christian idea that history is the divinely ordained process of man's salvation" (*Philosophy* 40). However, Wiebe's conception also offers an alternative understanding of humanity's place in the world, one that questions the liberating role of science and technology. As Žižek argues, science's "function is to provide certainty, to be a point of reference on which one can rely, and to provide hope (new technological inventions will help us against diseases, and so on)" (*In Defense* 446). Further, "The paradox effectively is that, today, science provides the security which was once guaranteed by religion, and, in a curious inversion, religion is one of the possible places from which one can develop critical doubts about contemporary society" (Žižek, *In Defense* 446). Thus, the angel reveals a lack in the dynamic Western culture, for which Christianity is an origin, and forces the reader to raise doubts about contemporary society.[21]

Rather than human dominance over nature, submission to an other determines what our relationship to the world should be: "The sacrifice acceptable to God is a broken spirit; a broken and contrite heart" (Psalms 51.17). David recognizes how he has abused his power and brought ruin on his neighbour. His recognition and subsequent confession are quite different than the mastery embodied in "the huge plant" in which "oil ran... gurgling in each precisely numbered pipe and jointure, sweet and clear like golden brown honey" ("Angel" 191) (see Figures 4 and 5). Joy Anne Fehr reads the superintendent's vision as pointing back to the geological origins of bitumen deposits and forward to a future in which bitumen extraction

FIGURE 4: *Suncor upgrader.* [*David Dodge, the Pembina Institute*]

FIGURE 5: *Syncrude plant.* [*Greenpeace/Ray Giguere*]

has led to "a veritable hell on earth" (214). Regardless of our opinion of the consequences of Christianity in North America, and without calling for some kind of "return to religion," Wiebe's story envisions something lost in the pursuit of human freedom, the freedom pursued through the precise analysis of numbered pipes. Fehr argues that the superintendent's "instinct is to revert to a familiar construction—Alberta is the land of milk and honey, a land of wealth and promise" (215): the comparison of oil to honey suggests the view of Alberta as the promised land, flowing with "milk and honey" (Exodus 3.8), but also makes an ironic comment about sustainability. Milk and honey are food sources produced with little or no human intervention; the promised land is flowing with these foods when people arrive there. By contrast, bitumen can only create human wealth through labour and analysis.

Alternative visions are necessary if we are to displace the displacements justified in the name of independence and mastery through progress, if we are to recognize the lyric coherence of which we are a part but do not control. Bott claims,

> Human ingenuity has already accomplished a great deal by making the oilsands economically competitive with conventional oil. Environmental and social challenges are being engaged. Continuous improvement in science, technology and management are helping to overcome the remaining challenges to meet society's expectations for sustainable development. (39)

The problem with "sustainable development" remains its continued orientation toward human needs and its assumption that these needs are knowable in an accountable way. Bott provides the standard definition of sustainable development: "development that meets today's needs without compromising the ability of future generations to meet their needs" (31).[22] Shiell and Busby define sustainability "to mean that a given policy can be continued at the current level indefinitely" (2). Both of these definitions indicate a view in which the Enlightenment promise of "the end of history" continues to be persuasive: we can solve the problem of scarcity on a worldwide level indefinitely (bitumen extraction will create the land of "milk and honey"). As Grant has it, "Modern human beings since their beginnings have been moved by the faith that the mastery of nature would lead to the overcoming of hunger and labour, disease and war on so widespread a scale that at last we could build the world-wide society of free and equal people" (*Technology and Justice* 15). However, while "one must never think about technological destiny without looking squarely at the justice in those hopes" (Grant, *Technology and Justice* 15), one must also consider the need for justice for the human and non-human nature commodified and sacrificed in the pursuit of those hopes. Lyric points to this alternative call for justice; Bertha's choice in Wiebe's story points to this alternative.

Lyric is continually marginalized as an alternative by those who suggest it is comprised of "smoke and mirrors...just metaphors" (Zwicky, *Lyric* 326). Zwicky acknowledges that lyric moments, indeed, "are just metaphors" and explains, "If there is someone who truly has no idea what I mean by 'analytic style' or 'resonance,' then these remarks will have no meaning for that person" (*Lyric* 326). After his lyric vision, the superintendent cuts himself off from that moment of resonance and recognition of sinfulness and loss; he turns back to the oil, "sweet and clear like golden brown honey" (191) in a moment that may be read as analogous with Lougheed's claim to hear and understand Crowfoot. The oil compensates for the loss of the surface of the earth, for our inability to leave things alone.[23]

While I fear that oil executives and members of government, like Lougheed in Wiebe's play, may well have no idea what values exist beyond their rational-analytic understanding, or choose, like the superintendent, to deliberately turn away from such moments of recognition because of the comforts oil wealth can buy, I continue to search for signs that Albertans are looking for a new story around which concerns over ecological destruction can coalesce, around which a different story, a lyric story, can find resonance.

In the findings of "The Oil Sands Survey: Albertans' Values Regarding Oil Sands Development" (2008), Cambridge Strategies found that "Albertans see the ecological issues of wildlife, Greenhouse Gases [*sic*], water and reclamation being of very significant concern" (Chapman et al. 4), with 73 per cent believing water usage is determined by the needs of industry and 66 per cent believing the reclamation pace is set to serve industry needs (5). Since the publication of that survey, Cambridge Strategies co-chair Ken

Chapman went on to become the executive director of the Oil Sands Developers Group,[24] and argued for responsible corporate citizenship and the view that Wood Buffalo can be the greenest municipality in Canada because "we can afford it" (Chapman, Petrocultures). Satya Das, Cambridge Strategies' other co-chair in 2008, went on to publish *Green Oil*, which argues, similarly, that the profits from bitumen extraction can be used to transition to a green economy. The survey's conclusion that Albertans "are very clear in that they want a green lens to trump any economic or growth dynamics" (6) may have led to a change in rhetoric but industry continues to expand. In the years since Cambridge Strategies commissioned their survey, it does not seem that Albertans' values have changed. A 2014 survey by Ipsos Reid for the Alberta Energy Efficiency Alliance found that "76 per cent of Alberta respondents either support or strongly support the provincial government enacting stronger greenhouse gas rules on large emitters" (Thomson). At the same time, *Edmonton Journal* columnist Graham Thomson claims that "Alberta's oilsands are the fastest growing source of emissions in Canada and the government doesn't really know how it will stop those emissions from increasing." While this phrasing suggests it is the government's responsibility to control emissions, that responsibility operates in relation to individual consumers' desires and habits. Both respondents to a survey and candidates for public office can express wishes for rules on emissions and protection of the environment; however, when those wishes conflict with a deeply held cultural faith in the status quo narrative of growth, change is hard to come by.

For Zwicky, lyric cannot offer a positive political alternative (*Lyric* 514). But, in revealing the qualitative values that are marginalized by political narratives of progress, lyric can open us to possibilities for what one *should* do "beyond all bargains and without an alternative." As Zwicky explains, "there may even be such a thing as lyric community—though when more than one human is involved, the odds against its achievement increase dramatically with every individual added, and with time" (518). This does not mean that we should abandon politics—"To accept the need to assume political responsibility in civil society is a mark of domesticity" (*Lyric* 522), a mark of growing up, or what Angus calls "inhabitation"—but it does mean that we must engage in politics without the clarity of lyric, which is only momentary and context specific.

In "The Angel of the Tar Sands" the superintendent's final actions are a rejection of the revelation the story contains, but the story itself opens an alternative for readers. When the angel in the story first moves, the characters "staggered back, fell" (189), and "the superintendent found himself on his knees staring up at the shape which wasn't really very tall, it just seemed immensely broad and overwhelming" (189). The angel interrupts the status quo, the liquidation of bitumen, the superintendent's commitment to keep the plant running at "capacity" (188), to "shut up the environmentalists" (188). The angel's completely unexpected appearance suggests that the land is more than the characters have yet imagined.[25] The public outcry over the duck deaths, though only momentary, indicates a similar recognition. Seeing "the layers of strata...ripped away" and feeling the meaning of Psalms 51 (191) open a gap in the superintendent's initial assumptions, re-present them, in Spivak's sense. Although the superintendent rejects this alternative in the end, if readers listen to the story and attend to the meaning that the superintendent "could not have explained" (191), but that is acted upon by the character Bertha, then we can glimpse an alternative beyond the realm of rhetorical representations. We need not be satisfied with giving up what we owe to others, and to otherness—that indefinable, absolute, uncommodifiable, useless alterity, whose best name is wilderness—*if* we make enough money in the trade.

2 **Oil** Sacrifices

Petroculture and (E)utopian Imaginings of Progress

THERE IS A NARRATIVE OF PROGRESS built into the political and economic culture of Canada. The surmounting of obstacles has been, and remains, prominent in narratives of settlement and development, which can be located within a larger narrative of Western rational, scientific, and technological progress. As Karl Clark, the University of Alberta scientist who pioneered the hot water extraction process for separating oil from sand, wrote in 1949, "The seamed cliffs on the Athabasca carry a smirk on their faces, and no one who visits the Northland can miss its meaning. It says plainly: 'You are doing big things with everything else in this North Country, but I have still got you beat. When are you going to do something about it?'" ("The Athabasca"). Nature poses a challenge; human ingenuity holds the answer. When Clark wrote those words, something was being done about it: attempts to commodify bitumen had been ongoing for nearly seventy-five years. It was only toward the end of his career that those efforts started to show success. Such efforts, though, require the sacrifice of various forms of human and non-human community, sacrifices justified in the name of progress.

Pieter Vermeulen shows how both Benedict Anderson's and Jean-Luc Nancy's understandings of community find common ground "where they reflect on the role of death and loss in the constitution of community" (96). It is in the community's response to loss that it justifies its existence. Perhaps we can see this type of response most clearly in something like social contract theory as articulated by Thomas Hobbes, where the state arises when individuals agree to a mechanism for mediating conflicts between their interests: if I am wronged, if my property is stolen, or I am murdered, the community, the state, or the sovereign will prove its legitimacy in how it responds. Gismondi and Davidson argue, "State legitimacy authorizes levels of control over the social order that no other institution enjoys. The legitimacy of a state thus includes a concession of the right to use force, albeit within limits that are not always clear or fixed. The justification for such authority is offered in exchange for social welfare, security, protection of individual liberties, and of course stewardship of natural resources and environmental well-being" (171). However, an extension of this contractual function operates through a symbolic process of abstraction, which seeks to enable mourning for members of the community who die in its service. Part of industry's challenge in developing bitumen is creating methods for managing the suffering produced as a result of human and non-human sacrifices required to produce that development. The rhetorical battle over bitumen development (among various political parties, federal and provincial governments, industry, academics, NGOs, environmentalists, and so on) can be understood, in part, by considering who is attempting to manage the losses produced by bitumen development and who is trying to mobilize them to change the status quo.

Literature is one site of response, considered by both Anderson and Nancy, that can justify the forms of loss that enable community or counter them by imagining other ways of responding.[1] This chapter will think about a few of the forms of loss caused by oil capital in Alberta, consider how government and industry responses attempt to put those losses in the service of liberal capital, and explore the possibility for literature to defy those attempts by reading Charles Mair's "The Last Bison," Sidney Ells's verse epilogue and epigraph to his *Recollections*, Mari-Lou Rowley's "In the Tar Sands, Going Down," and Thomas King's "Alberta Oil Sands Land."

Ducks and the Nation-State

Perhaps the most widely publicized loss to bitumen production was the deaths of 1,600 ducks on Syncrude's Aurora tailings ponds in April 2008. Even though there have been subsequent losses caused by bitumen development, these ducks continue to haunt the industry and government in ways that the injury and death of workers and cancer rates in First Nations communities downstream do not (see Christian, "Killed Workers" and Brooymans, "Fort Chipewyan" A1).[2] Despite the continuing recurrence of this scene of loss, there has not been a response to the duck deaths that has successfully changed the social licence for development. Partly this is due, I will argue, to the primary role of the nation-state as mediator and diffuser of tensions, even when it has ceased to be the dominant imagined community at work. As Len Findlay has put it,

> transnational corporations and the United States both need nation-states as satellites and proxies to talk (down) to, and to help develop the deregulating political and fiscal instruments necessary for further "globalization" of the market economy; at the same time, citizens need nation-states as sites of resistance to such political and economic hegemony exercised by unelected elites and insiders. (298)

In addition to serving as an actual site of resistance, the nation-state also serves as a perceived site of resistance, a focus for citizen action, in situations where its role is limited.

Where the potential for effective resistance does not exist, or such change would be counter to dominant interests, the nation-state provides a focus for symbolic action. Vermeulen continues, "The nation-state relies on the readiness of its subjects to die for their community, and in order to promote such self-sacrifice, it forecloses the horror and sadness of dying by integrating death with an overarching framework in which it acquires public value and meaning" (97). This logic of death and self-sacrifice seems to operate at levels of imagined community other than the nation-state as well: as just one example, despite the deaths of and injuries to workers in bitumen operations, extraction continues; workers continue to put themselves at risk in the service of capital. To understand how the logic of self-sacrifice underwrites bitumen development we need to rethink Benedict Anderson's foundational theorizing of the imagined community. We can see, through the public relations work of the bitumen industry, that, despite negative press, it has managed to retain its social licence to operate. That pollution generated by industry is threatening human and non-human lives is both denied and managed within the myth of progress. The illness and death of workers, Natives, and wildlife is both denied and shown to be compensated for by the benefits of development—energy security, profit, growth, jobs, funding for culture, education, and so forth. Gismondi and Davidson, for example, describe how "critical analysts label such rhetorical techniques as 'bridging effects,' that is, ways of speaking that ties tar sands oil (which has low social standing) to Canadian health-care (a high prestige discourse)" (104). In other words, taxes and royalties paid by bitumen companies pays for health care, so if you like health care you can't be against bitumen extraction. The decision to develop bitumen is a trade-off, and the global considerations trump the local; the urban trump the rural; the many trump the few. Therefore, even thoughtful commentators on bitumen development such as Peter Silverstone, vice-chair of the board of the Edmonton Economic Development Corporation, and Andrew Leach, University of Alberta business professor, often focus on greenhouse gas emissions rather than heavy metals or polycyclic aromatic hydrocarbons, for example, because one is a "global" problem while the other merely local (Silverstone 3; Leach, "Don't Let EU" A15). An additional variant of this response is pragmatic relativism: Ezra Levant makes this point explicitly in *Ethical Oil*, where he argues—on what he claims are progressive grounds—that Alberta bitumen is better than oil from other nation-states: better environmentally, better in its contribution to world peace, better in its treatment of minorities, and better in terms of economic justice.[3]

When these forms of management fail, though, to foreclose the horror of particular losses, as with the death of the ducks or revelations of water pollution caused by industry by University of Alberta water experts Erin Kelly and David Schindler or Environment Canada scientist Derek Muir and Queen's University biologist John P. Smol, promises of improvement fit into the Enlightenment narrative that the application of reason will achieve its liberating potential sometime in the future. Syncrude will improve its duck deterrent systems; the provincial government will convene a panel of scientists to compare the water data and make the appropriate changes. After the duck deaths, then Syncrude CEO Tom Katinas made a "promise to do better" ("An Apology from Syncrude"). This, as we will see, is the promise of modernity—the future will be better; any deaths endured today will have been worthwhile because of the kind of society that we will inhabit in the future. In the subsequent court case, in which Syncrude was charged under provincial environmental laws and the federal Migratory Birds Convention Act, the company was fined $3.2 million, despite their defence lawyer's claim that "If...Syncrude is guilty of this crime, the government is complicit and the industry is doomed" because "If by having a tailings pond we're guilty of this charge, we have to stop having tailings ponds," and if that's the case, "we're done" (Henton A1). By implication, if industry is doomed, then so is the future that it is supposed to make possible. Of course, the guilty verdict and the fine have not doomed the industry or put an end to tailings ponds, despite the truth in the lawyer's claim about government complicity. If it is against the law to emit a deleterious substance into the environment, and the government has granted a licence to do precisely that, then the government is complicit and the verdict should require ending the use of tailings ponds. The verdict and the fine, rather, have enabled a return to the status quo by providing the appearance of justice being done and a form of symbolic closure to the sadness expressed over the ducks' deaths. The form the fines took—with some of the money directed to a research project into bird migration and deterrent systems, the Alberta Conservation Association, and a wildlife management program at Fort McMurray's Keyano College (Wingrove, "Syncrude")—can be read as an attempt at "integrating death with an overarching framework in which it acquires public value and meaning" (Vermeulen 97). Like soldiers dying for their country whose sacrifices are said to make the lives of their fellow citizens safer, the deaths of these ducks will make the world safer for other ducks and will make the coexistence of ducks and bitumen mining possible.[4]

However, as Imre Szeman puts it, "It is not that we can't name or describe, anticipate or chart the end of oil and the consequences for nature and humanity. It is rather that because these discourses are unable to mobilize or produce any response to a disaster we know is a direct result of the law of capitalism—limitless accumulation—it is easy to see that nature will end before capital" ("System" 820). The question, then, is why are these critical discourses unable to mobilize any effective response? Why does pointing out the flawed logic of an imagined community not lead to any collective reimagining?

Spinning Sacrifice

For Anderson, "the national imagining of community, when considered as a particular style, in the final analysis consists in a particular stylization of death" (Vermeulen 99). Thus, "imagining community differently" Vermeulen concludes, "also means finding another way to relate to the death of the other" (104). We might then understand the current debate surrounding bitumen development as a tension in what has been the dominant stylization of death, loss, and sacrifice required by development. Within the collective imaginary there continues to be a series of attempts to find other ways of relating to the death, loss, and sacrifice of the other, an other understood as both human and non-human. How do we make sense of the duck deaths? What story do we tell? Is it part of a larger ecological tragedy or a freak accident, part of a modern system of slavery,[5] or, *pace* Levant, better than the alternative(s)? Could the story change the way we develop bitumen?

The way we imagine collectivities in time is crucial in this respect. This imagining goes beyond naming, describing, anticipating, or charting effects. For Anderson, national imagining takes place in "homogenous, empty time," where "simultaneity is, as it were, transverse, cross-time, marked not by prefiguring and fulfillment, but by temporal coincidence, and measured by clock and calendar" (24). Thus, though "an American will never meet, or even know the names of more than a handful of his

240,000-odd fellow-Americans...he has complete confidence in their steady, anonymous, simultaneous activity" (26). Today, though, while this form of imagining continues to bind members of national collectivities, another form of imagining is, perhaps, even more important, and has shifted the focus onto a future that makes up for the failures of the past. Therefore, while Anderson points out that "having to 'have already forgotten' tragedies of which one needs unceasingly to be 'reminded' turns out to be a characteristic device in the later construction of national genealogies" (201), in moving beyond the nation as the key site of collective imagining, past tragedies are incorporated into the narrative as developmental stages. They need to be remembered as moments enabling progress. Anderson writes,

> English history books offer the diverting spectacle of a great Founding Father whom every schoolchild is taught to call William the Conqueror. The same child is not informed that William spoke no English, indeed could not have done so, since the English language did not exist in his epoch; nor is he told "Conqueror of what?". For the only intelligible modern answer would have to be "Conqueror of the English," which would turn the old Norman predator into a more successful precursor to Napoleon and Hitler. Hence "the Conqueror" operates as...[a] kind of ellipsis...to remind one of something which it is immediately obligatory to forget. Norman William and Saxon Harold thus meet on the battlefield of Hastings, if not as dancing partners, at least as brothers. (201)

While these moments might prompt critical reflection on the role of the nation-state and make one question identifying with it, recalling them does not necessarily enable resistance to the narrative of modernity. This narrative can assimilate them as stages in the ineluctable march of abstract progress.

A *Fort McMurray Today* article titled "Killed Workers Remembered" recounts the observation of an International Day of Mourning for workers killed and injured on the job, and quotes Kirk Bailey, vice-president of Suncor, as saying, "We can do better and we are going to commit ourselves to having a workplace that's absolutely free of injuries and I know we believe that at Suncor and I know the other industry players here in the region are equally committed to that goal" (Christian). This focus helps to defer potential opposition to worker's deaths, or bitumen extraction more generally, into the future (for example, we understand we are not perfect, but we're working to improve and we will not settle for less than perfection *at some unnamed future date*).[6] Similarly, industry and its apologists like to trot out statistics such as the fact that "to produce one barrel of oil sands oil takes 38 percent less emissions now than it did in 1990" (Levant, *Ethical Oil* 6). Yes, worker deaths are tragic; yes, pollution is bad; but these sacrifices are necessary to secure the (e)utopian[7] future of limitless energy free of negative effects.

Writing about tombs of unknown soldiers, Anderson states,

> The cultural significance of [tombs of unknown soldiers] becomes even clearer if one tries to imagine, say, a Tomb of the Unknown Marxist or a cenotaph for fallen Liberals. Is a sense of absurdity unavoidable? The reason is that neither Marxism nor Liberalism are much concerned with death and immortality. If the nationalist imagining is so concerned, this suggests a strong affinity with religious imaginings. (10)

While the monument in Fort McMurray's J. Howard Pew Park is not a tomb to an unknown soldier, Carol Christian's photo shows a ceremony strikingly similar to the ritual laying of wreaths at cenotaphs on Remembrance Day (see Figure 6). Perhaps the focus is not on death and immortality in the way Remembrance Day ceremonies are, not on the deaths that secured the immortality of the nation. But how is death being imagined here? Is the labour movement concerned with death and immortality, with religious imaginings?

The International Day of Mourning was established by the Canadian Labour Congress in 1984 as a national day of mourning for workers killed or injured on the job. The quote from the representative of labour provides a striking contrast to that of the industry representative, Kirk Bailey. Christian quotes Perry Turton of the International Brotherhood of Electrical Workers as observing, "We often hear the numbers of people that have passed through the years...while at work. What we won't hear is the

FIGURE 6: *An unidentified man lays flowers at a memorial for killed and injured workers in J. Howard Pew Park in Waterways.* [Carol Christian, Fort McMurray Today]

names of those that have passed." He continues, "This is something that bothers me as it feels we are allowing industry to turn us all into statistics. We are not numbers. We are people with families, friends" (Christian). The focus on numbers rather than names abstracts from the particularity of individual suffering and death and shifts the discussion from mourning to progress, a shift that is also the focus of environmental public relations.

The Government of Alberta's *Oil Sands Information Portal* includes the following statement under the heading "Responsible Development" in the first section "About the Oil Sands": "Clean energy is one of the biggest challenges of our time. Overcoming the obstacles will require efforts from all of us around the world. We'll have to change the way we think about and use energy, alter our consumption patterns, and consider our responsibilities in a global context." Although this statement does not include any specific references to the use of fossil fuels, let alone bitumen, in overcoming the obstacles to clean energy, the context seems to imply that the use of bitumen does have a role in the process of transformation being described. The remainder of the screen does not mention clean energy again, going on rather to discuss regulations and land-use frameworks; the first section frames what follows to imply that sufficient regulation can, somehow, turn bitumen into clean energy.

We might read this as a defence of the status quo through a promise for the future. Clean energy will exist someday, sufficient regulation today will ensure that there is some natural environment left when it does. Meanwhile, resource industries, enabled by government and citizen complicity, sacrifice the environment and society in the name of economic competition. Downstream from Fort McMurray, Fort Chipewyan, the oldest continual settlement in Alberta, provides one example of sacrifice. For many residents, hunting, trapping, and fishing remain tied to an intimate knowledge of place and natural cycles. When production continues twenty-four hours a day, three hundred and sixty-five days a year, bitumen's industrial activity of extracting abstract capital from concrete places is incompatible with such a relationship to place. Not only is their *way of life* being sacrificed for capital, but, given the high rates of unusual cancers that are appearing in the community, *lives* are also being sacrificed. Gismondi and Davidson note, "One particularly useful means of maintaining legitimacy in this case would be to distract attention away from clear breaches in instrumental legitimacy, an increase in cancer rates downstream from the tar sands, let us say, by drawing attention to the moral necessity of tar sands development in order to curb terrorism, or provide jobs that feed families" (172). Allan Stoekl reads Georges Bataille via Heidegger to come to the conclusion, "modern subjectivity—subjectivity that itself can be objectified, quantified—is inseparable from an instrumental conception of 'resources': matter is now quantity—measured, hoarded, and then spent. The self is a function of the world as standing reserve, the collection-disposal of accumulated raw material. And the self becomes raw material as well" (*Bataille's Peak* 139). This view of subjectivity as object, of the self as raw material, suggests that the language of sacrifice is foreign here. Indeed, the language of sacrifice is a way of resisting this objectifying view of otherness.

An example of how literature can point beyond the logic of liberal progress resulted from Thomas King's unsuccessful run for the NDP in Guelph in the 2008 federal election; as part of his campaign, King wrote a series of short stories attacking Stephen Harper and Stéphane Dion, then leader of the Liberal Party of Canada. One of these sees King's credulous narrator persuaded to change his plans for vacationing in the Rockies to tag along with the two federal leaders to visit "a new, all-inclusive resort near Fort McMurray called Alberta Oil Sands Land" because, according to the prime minister, "nature is overrated." After describing the various attractions, such as "the Ralph Klein Environmental Pavilion," unfortunately closed for repairs; "the Ralph Klein Waste Water Park," which "looked, for all the world, like a series of interconnected toxic-waste holding ponds right on the Athabasca River"; "the Ralph Klein Pollution Pavilion"; "the Ralph Klein Musical Effluent Fountain"; and "the Ralph Klein Earth Mover rides," King writes, "the best part of the entire vacation was watching the sun set each evening on what was left of the Athabasca River. Mr. Harper would strum his guitar and tell us how his heart swelled when he looked out over the landscape and saw the sheer beauty that human ingenuity and corporate genius could create" and "Even Mr. Dion got a little teary as we all watched the smoke stacks discharge their billows of pollutants. He said that the soft clouds floating over the high prairies reminded him of the old days when Native people sat around their council fires and told stories about living in harmony with the earth." Aside from the less-than-subtle political attacks on his opponents and the repeated invocation of the ghost of Ralph Klein, sure to strike fear into the hearts of central Canadians, this story, like Wiebe's, makes a spiritual point that is missing from, and perhaps unspeakable in, government and industry discourse. As the 2007 report of the Standing Committee on Natural Resources, *The Oil Sands: Toward Sustainable Development*, has it, "While Aboriginal peoples in the Wood Buffalo region are deriving some economic benefits from oil sands activities, it is not clear whether such benefits *compensate for* the long-term social and environmental impacts of this very rapid industrialization" (Richardson 50, emphasis added). This statement presupposes that some degree of economic benefit *could* compensate for the industrialization and pollution of traditional lands, high rates of virulent cancers, and rapid changes to preexisting human and non-human communities. Or, as one of the committee's recommendations implies, if "the federal government take[s] necessary measures to improve living conditions for the aboriginal communities," then the loss of a way of life will be acceptable. This compensation is seen, for example, in the "nearly $2 million" Syncrude "invested" in "Aboriginal community projects" from 2010 to 2011 (*Are the Oil Sands* 76). King indicates, through the irony of having Dion "remember" pre-colonial Native culture, both that something of more than quantifiable economic value is lost when humans assert their independence through dominance over nature and that the "image of the indigene" (Goldie) is still used very often as a symbol for non-Native desires and ideas about authenticity. The beauty of a sunset can still invoke the idea of natural harmony for the sentimental, despite evidence to the contrary. As Slavoj Žižek has written, the problem with confronting ecological catastrophe

> resides in the unreliability of our common sense itself which, habituated as it is to our ordinary life-world, finds it difficult to really accept that the flow of everyday reality can be perturbed. Our attitude here is the fetishistic split: "I know very well (that global warming is a threat to the entire humanity), but nonetheless...(I cannot really believe it). It is enough to see the natural world to which my mind is connected: green

> grass and trees, the sighing of the breeze, the rising of the sun...can one really imagine that all this will be disturbed? You talk about the ozone hole—but no matter how much I look into the sky, I don't see it—all I see is the sky, blue or grey!" (445)

The story of sustainable development serves to placate us as we go about our lives and see the sun rise and set as it always has. It enables us to avoid facing the violence required by bitumen development. The idea persists that the oil sands can be part of a "clean energy future" (Richardson 53) as industry makes "progress toward sustainable oil sands development" (Syncrude, *Synergy: 2008/09 Sustainability Report* 2). "Alberta Oil Sands Land," on the other hand, suggests that the pace of development in the Athabasca region creates uninhabitable futures, and therefore unsustainable ones, in the etymological sense of a land incapable of producing food for human sustenance, even as it finances the present. Development creates debits that cannot be compensated for with a two-million-dollar credit.

However, each apparent failure of science and technology to produce the promised perfection in our world creates an opportunity for reimagining that collective myth. Vermeulen states, "Nancy's notion of the inoperative community receives its critical momentum precisely by resisting that process of abstraction [by which individual deaths are seen as glorifying a collectivity]. Against such ideological abstractions of death, Nancy puts forward a notion of existence as 'unsacrificeable'" (97). Similarly, we might argue that current deaths are not justified by the promise of fewer, or no, deaths in the future. Those individual workers with families and friends, those members of the communities downstream from bitumen mining operations, cannot be sacrificed; more money, better schools, well-funded cultural institutions, amusement parks do not compensate for those deaths: nothing does.

While the imagined community denies the irreducible singularity of death by glorifying and monumentalizing it, or ignoring it, or projecting it into a (e)utopian future, or rationalizing it as being in the common good, Nancy considers death and existence as "always inappropriable and always there, each and every time present as inimitable" (Vermeulen 97). This applies, I believe, not just to literal human death, but also to the ways in which other forms of loss are managed through abstraction.

Levant, with his publication of *Ethical Oil*, has shifted his argument from economic grounds to comparing the consequences of bitumen extraction with other forms of oil on what he calls "liberal" (Levant and Nikiforuk) grounds. If fewer people are killed by the bitumen industry in Northern Alberta than in Iraq, for example, then bitumen should be extracted faster, in Levant's view, to lessen reliance on Iraqi oil, and, thereby, save lives. He states, to illustrate this point, that every barrel of oil from Sudan, for example, "has a thumb or an eyeball in it"; there is, he states, "no comparison between the oil sands, which killed a thousand ducks...and 300,000 dead in Darfur" (Levant and Nikiforuk). Of course, this ignores the fact that demand equals supply, and Iraqi, Sudanese, Saudi, Libyan, Russian, and Albertan oil will all find buyers, and, in fact, are indistinguishable (i.e., fungible) in the world market. On this point Gismondi and Davidson argue that, in fact, "the Athabasca is land-locked, and thus the synthetic fuel is non-fungible—since it must travel by pipeline, it does not get sold 'on the open market,' it gets sold to whoever is on the other end of the pipe" (161), which helps explain the Alberta and Canadian governments' urgency to build the Northern Gateway and Keystone XL pipelines. However, it does not mean that increase in bitumen production will reduce demand for oil from elsewhere; rather, it will drive down prices, meaning bitumen is sold at a greater discount compared to light oil from elsewhere. Gismondi and Davidson later point out that, "because staples commodities are virtually interchangeable—the performance of gasoline derived from Alberta tar is indistinguishable from that derived from Venezuelan oil to the average consumer—producing countries face a perpetual buyer's market. Exporters are thus vulnerable to reputational threats that may lead to consumer boycotts and investor hesitancy" (173). Therefore, it is impossible for bitumen production to accelerate fast enough to reduce, let alone eliminate, demand for oil from elsewhere. Gismondi and Davidson note, "considering that supply and demand are in a dialectic relationship, this trope [of the need to increase production to meet increasing global demand] becomes a self-fulfilling prophecy, since increases in supply availability lead to lowered prices, which in turn lead to increases in consumption" (156). More importantly, though, this kind of relativist calculus cannot account for Nancy's point about the unsacrificeable. Each death is full of horror and sadness, and this is true of duck deaths as well as human (as the published photographs and videos, and international reaction to the footage of the ducks as they were dying made evident). An ethical ranking of types of oil does not absolve us of culpability with the problems of even the best type.[8]

I wouldn't go so far as to suggest that death is never justified; it doesn't take long to recognize the multitude of deaths that make my life and lifestyle possible (including the duck deaths). In *The Gift of Death*, Jacques Derrida states, "I am responsible to any one (that is to say to any other) only by failing in my responsibilities to all the others, to the ethical or political generality. And I can never justify this sacrifice, I must always hold my peace about it. Whether I want to or not, I can never justify the fact that I prefer or sacrifice any one (any other) to the other" (70).[9] Levant does not hold his peace about this. He justifies it. Suncor Vice-President Kirk Bailey does not hold his peace about this. He justifies the death of a worker today—he sacrifices that worker—for the imagined life of a worker tomorrow, for the future (e)utopia where "hunger and labour, disease and war" have been overcome (Grant, *Technology and Justice* 15), for the "end of history." Indeed, in writing this chapter, I have been unable to hold my peace about it. We might even say of Derrida, as Blanchot did of Wittgenstein, "Wittgenstein's all too famous and all too often repeated precept, 'Whereof one cannot speak, there one must be silent'—given that by enunciating it he has not been able to impose silence on himself—does indicate that in the final analysis one has to talk in order to remain silent" (Blanchot 56). As I hope to suggest at the end of this chapter, although it can only do so temporarily, literature provides one means of talking in order to remain silent, enables us to dwell in our failure to all the other others, allows us to recognize and hold our peace about it.

Lament for the Nation's Petroculture

Indeed the justification of others' sacrifice may be the narrative of modernity in general, of which nationalism is only a reaction and mediation, the drawing of a border, a limit. Anderson's famous definition of a nation as "an imagined political community—and imagined as both inherently limited and sovereign" (6) is useful here. He clarifies the term "limited" by saying,

> The nation is imagined as limited because even the largest of them, encompassing perhaps a billion living human beings, has finite, if elastic, boundaries, beyond which lie other nations. No nation imagines itself coterminous with mankind. The most messianic nationalists do not dream of a day when all the members of the human race will join their nation in the way that it was possible, in certain epochs, for, say, Christians to dream of a wholly Christian planet. (7)

However, it is not impossible to imagine an entirely capitalist (or communist) planet. Indeed, to return to the Canadian context, in *Lament for a Nation* George Grant argues against the progressivist view of history, claiming that the necessary and the good need not be equated, that because the death of one individual or group may be necessary it is not necessary to claim it is good: the necessary becomes equated with the good, he writes, "because men assume in the age of progress that the broad movement of history is upward. Taken as a whole, what is bound to happen is bound also to be good. But this assumption is not self-evident. The fact that events happen does not imply that they are good. We understand this in the small events of personal life. We only forget it in the large events when we worship the future" (37). As discussed in the Introduction, there is a bias towards such upward movement, which makes upward counterfactual claims more persuasive than downward counterfactuals, regardless of which side the "facts" support. Grant goes on, in the next chapter, to claim,

> The universal and homogeneous state is the pinnacle of political striving. "Universal" implies a world-wide state, which would eliminate the curse of war among nations; "homogeneous" means that all men [*sic*] would be equal, and war among classes would be eliminated. The masses and the philosophers have both agreed that this universal and egalitarian society is the goal of historical striving. It gives content to the rhetoric of both Communists and capitalists. This state will be achieved by means of modern science—a science that leads to the conquest of nature. (52)

Gismondi and Davidson describe the visual representation of the Great Canadian Oil Sands (GCOS) plant this way: "the visual impact of professionally engineered production plants on the edge of the boreal wilderness invoked paradigms of progress and human domination over nature more than words ever could, especially when contrasted with early images of boiling pots of oil and slipshod systems of mining and separation" (60). The continual references to science and technology in government and industry documents about bitumen extraction echo Grant's point, as does the funding of two post-secondary programs with the Syncrude fine.[10] For Grant, "It is the very signature of modern man to deny reality to any conception of good that imposes limits on freedom" (*Lament* 55). Therefore, insofar as nations are limited, they are counter to the thrust of historical progress, and, in the case of Canada, in Grant's view, fated to be swallowed up by the modern liberal capitalism of the US empire. For Grant, this event took place in 1963 with the defence election, in which Lester Pearson defeated John Diefenbaker, allowing his minority government to arm Bomarc missiles on Canadian soil with nuclear weapons.

I would like to link the end of Canada as a nation to different but historically contemporaneous events. The publication of the Blair Report in 1950 showed that bitumen could be extracted and upgraded at a profit (produced for $3.10 and sold for $3.50 a barrel) (K. Clark, "Commercial Development" 5). This led, in 1953, to the creation of the Great Canadian Oil Sands consortium, whose major investor was Sun Oil of Philadelphia.[11] In *Lament for a Nation*, Grant asks if his readers can "imagine Canadians expropriating the oil properties and taking on international capitalism as Cardenas did [in Mexico] in the 1930s?" (45). While discussion of nationalized development of bitumen did occur in Canada at the time, Karl Clark expressed the dominant view about it in a letter of June 12, 1963: "The political opposition parties in Alberta are saying that the Social Credit government is selling the oil sands, or rather giving them to the Americans—that Alberta should be doing the developing itself. I think Premier Manning is very willing to let private enterprise take the risk of a first plant, at least." This willingness to cede to American capital the right to extract Canadian bitumen, the right to abstract the particularities of the Athabasca region into money, has proven more significant than the arming of Bomarc missiles with nuclear warheads. The particular losses suffered in the Athabasca region may prove to be less widespread and catastrophic than those of a nuclear war would have been (and than widespread nuclear energy use could still be, as the events following the March 11, 2011 tsunami and the Fukushima Daiichi reactor have intimated again), but they still have a singular horror.

In a letter of July 14, 1960, Clark wrote that at hearings in Calgary regarding the GCOS application for its first large-scale extraction plant (45,000 barrels a day), Bob Hardy testified with "quite a story about how we were going to have to stack tailings up 160 feet and were in all likelihood going to start slides that would engulf everything in catastrophic ruin. That picture lost its horror when it was pointed out that our tailings were going back into the mined-out pit enclosed on all sides." However, the horror returned for Clark in 1965 when he visited the GCOS site. His daughter Mary writes, "It affected him deeply to see the landscape of his beloved Athabasca country scarred as the construction gangs began stripping away the overburden for the operation" (Sheppard 89), and, shortly before his death, he confided that "he had no wish to return again to the scene" (90). I can easily imagine Clark lamenting the loss of a particular place, dying with the irony that his life's work had helped it happen. Grant notes that "the practical men who call themselves conservatives must commit themselves to a science that leads to the conquest of nature. This science produces such a dynamic society that it is impossible to conserve anything for long" (*Lament* 65). I don't know if Clark called himself a conservative or considered himself one in any sense. He voted Liberal federally and against the Social Credit dynasty provincially (not, he claimed, in a letter of June 12, 1963, "that [he] want[ed] to see the Social Credit government defeated but rather that [he] would like to see the government have a decent opposition in the Legislature"). Regardless, at least at the end of his life, the scene described by his daughter shows a man who wanted to conserve something. Karl Clark saw the intrinsic value of the northern boreal forest sacrificed, and he turned away. Today, the rhetorical yoking of sustainability to progress, in the sense of "advancement to a further or higher stage...growth; development, usually to a better state or condition; improvement" ("Progress"), provides a convenient

FIGURE 7: Syncrude's Wood Bison Gateway stands at the beginning of the Matcheetawin Discovery Trails (Cree for "beginning place") and the Sagow Pematosowin (Cree for "living in peaceful coexistence with the land") thirty kilometres north of Fort McMurray. [Syncrude]

excuse for doing the same, for not facing the negative consequences of human ingenuity and corporate genius or trusting that our intelligence will solve the problems it is currently creating.

Bitumen Conserving Bison

The management of the form of ecological loss that affected Clark so deeply near the end of his life is undertaken in many ways, but is perhaps seen most clearly in the representation of bison. They are monumentalized in Syncrude's Wood Bison Gateway, which marks the entrance to the Syncrude site, and Wood Bison Viewpoint, which concludes both the Suncor and Syncrude tours, by offering tourists a chance to see wood bison grazing on "reclaimed" land.

In its 2007 sustainability report, Syncrude describes how it "established a small bison herd on reclaimed land in 1993" (*Canadians Want* 35). Since then the herd has expanded and the company has for several years "assisted the Fort McKay First Nation in arranging a traditional bison harvest" (35).

FIGURE 8: Bison grazing on reclaimed land at Wood Bison Viewpoint, forty kilometres north of Fort McMurray. [Syncrude]

As Nicole Shukin argues in writing about her tour of Syncrude's operations, Syncrude "pursues what Mark Seltzer calls a '*logic of equivalence*' between the space of industrial extraction and that of cultural preservation" ("Animal" 139). It is *because of* bitumen mining that the Fort McKay First Nation can conduct a "traditional bison harvest" with twelve animals that were "donated to the community from the Beaver Creek Wood Bison Ranch" (Syncrude, *Synergy: 2008/09 Sustainability Report* 35). As the word "donated" in the quotation suggests, however, the bison are the property of Syncrude until the corporation decides to give them to the community (Shukin, "Animal" 195). Thus, the bison exist within the realm of capitalist accumulation rather than as part of the wilderness commons they previously inhabited. Whatever community is possible between humans and bison is closely managed and controlled. The "bison harvest," a phrase that shifts the traditional Native economy into the agricultural realm, only occurs because of the benevolence of Syncrude's donation. We may ask why a "harvest" and not a "hunt"? For one, hunting is unpredictable and requires wild animals.[12] Harvests are supposed to be more predictable in terms of what they yield. Wild ducks, for instance, do not yield foie gras, as Levant suggests. The discussion of the bison appears in the section of Syncrude's sustainability report entitled "Aboriginal Relations," alongside information on the percentage of Natives in their workforce (8.5%), the amount of trade done with Native businesses ($143 million), investments in Aboriginal community projects ($1.2 million), and their involvement in Native education programs. Syncrude, as "the nation's largest industrial employer of Aboriginal people" (Shukin, "Animal" 134), shows itself to be responsible for the preservation of First Nations' culture in the area, but works to determine which aspects of the culture will survive and to ensure they are compatible with industrial development. Members of the Fort McKay First Nation actually need to book in as tourists to visit the bison (Shukin, "Animal" 195). Indeed, residents of Fort McKay may have preferred a moose project. As Hugh Gibbins says, "As a result of the 100-year absence of wood bison...and a strong familiarity among Fort McKay community members with moose hunting and processing, wood bison are not singled out in day-to-day or seasonal spiritual and religious ceremonies and practices" (qtd. in Shukin, "Animal" 196). The bison appear in Syncrude's sustainability report again in the section titled "Reclamation," where we learn that the herd "continues to thrive" on "700 hectares of reclaimed land" (53). That these hectares have not been certified as reclaimed is not mentioned: the bison serve as the only necessary evidence of reclamation. Further, because of the associations of First Nations' culture and bison in the cultural imaginary, which Syncrude has strategically exploited, their use of bison as a monument and as a "mascot" for the industry (Shukin, "Animal" 142) establishes the idea that bison and Native culture can be preserved and recovered under the direction and control of transnational capital. Nothing need be sacrificed to extract bitumen. The conflict between industrial development and wilderness preservation in the Athabasca region goes back at least to the signing of Treaty 8.

Charles Mair, who has been called the "Kipling of Canada" (Leonard xv), is a complex and controversial figure in Canadian literature and history, best known for his closet drama *Tecumseh*, his role in promoting Canadian westward expansion, and his arrest by Riel's provisional government during the Red River Rebellion. In *Through the Mackenzie Basin: An Account of the Signing of Treaty No. 8 and the Scrip Commission, 1899*, a commission for which Mair served as secretary, his conflicted relationship with Natives, nature, and progress continues. He notes, for example, that the purpose of the treaty is "to clear the way for the incoming tide of settlement...[which] is rapidly approaching, and when it comes the primitive life...will pass away forever" (7). He also records, without comment or qualification, Father Lacombe's declaration to the assembled inhabitants of the region at the treaty negotiations: "Your forest and river life will not be changed by the Treaty" (63). Like many explorers before him, he also notes that "in the neighbourhood of McMurray there are several tar-wells...[which will be] a substance of great economic value...beyond all doubt, and, when the hour of development comes...[they] will, I believe, prove to be one of the wonders of Northern Canada" (121). Indeed, the purpose of the trip was, he says, to place a region "rich in economic resources...at the disposal of the Canadian people" (22). This great economic potential has governed non-indigenous interactions with bitumen ever since. The contradictions revealed in Mair's writing—as both apologist for colonial expansion and industrial development and student of Native culture and advocate of wilderness preservation—strike me as symptomatic in the history of bitumen development, as those who pioneered its extraction also often express the desire to preserve the land *as* they encountered it before their work of industrial resource extraction was commercially successful.

At the time of Mair's visit, the wood bison was on the brink of extinction, as he well knew. He wrote, "Immediate steps should certainly be taken to punish and prevent poaching, or this band, the only really wild one on the continent, will soon be extinct" (97). Indeed, Mair's writing—in his poem "The Last Bison" and essay "The American Bison"—and his work for conservation were important in the creation of a bison preserve at Wainwright, Alberta (Braz; Leonard xxv). At the same time, though, Mair's poem tends to conflate Natives with the bison, and the death song of Mair's last bison is, in Katia Grubisic's words, "as much an account of an archetypal last Indian as it is the swan song of the buffalo. Both First Nations peoples and buffalo, in Mair's view, had gone from regal abundance to servile scarcity" (Grubisic). For Grubisic, "Mair's emotional and ideological investment in the myth of the ecological Indian seems to derive from a raceless wish for prelapsarian perfection; this prelapsarian wilderness, when harnessed by British civilization, promised the ideal balance between nature and morality."

Today, rather than British civilization, or, perhaps as its extension, the balance between nature and morality is harnessed to technological advance, and the myth of the ecological Indian, linked to the iconic wood bison, continues to serve a rhetorical function in the justification of developmental expansion as the means to preserving and accessing nature. In contrast to the upward counterfactual prophesy that closes Mair's "The Last Bison"—that one day the bison will return when—

> Naught but the vacant wilderness [will be] seen,
> And grassy mounds, where cities once had been.
>   The earth smile [*sic*] as of yore, the skies are bright,
> Wild cattle graze and bellow on the plain,
> And savage nations roam o'er native wilds again!
> (241)

Syncrude's wood bison remain firmly lodged within the narrative of progress even while the company makes use of their symbolic association with wilderness. It is worth noting that the conservationist movement arises in North America after the widespread bison slaughter of the late nineteenth century had, in many ways, solved the "Indian question" by breaking the economic and military strength of the Plains nations. Shukin writes, "no sooner was indigenous resistance to westward expansion violently emaciated than a sudden reversal in official and unofficial policy announced a new desire to protect bison" ("Animal" 179). Today, this desire to protect bison is seen in Syncrude's use of the herd in "a genetic preservation project that aims to create a sustainable future for the entire wood bison species" (Syncrude, *Canadians Want* 53). This time, it is sustainability that is possible *because of* bitumen extraction. Carolyn Merchant states, in the US context, "With Indians largely vanquished and removed to reservations by the 1890s, twentieth-century conservationists turned 'recovered' Indian homelands into parks, [and] set aside wilderness areas as people-free reserves where 'man himself is a visitor and does not remain'" (qtd. in Shukin, "Animal" 179). It is perhaps not coincidental, then, that the first reclamation certificate was issued to Syncrude in March 2008 for a 104-hectare parcel of land that is now a public park. Two possible "uses" of reclaimed land that the industry foresees are recreation and tourism.

Despite the assurances in the treaty negotiations that "Indians who take treaty will be just as free to hunt and fish all over as they are now" (Mair, *Through the Mackenzie* 58), and despite the fact that Natives have not disappeared from the area, contradicting the dominant expectation at the time the treaty was made, they continue to be pushed off the land, since much of the fish and game is unfit to eat due to pollution.[13] At the same time, those extracting capital from the land continue to construct themselves as benevolent parents employing, educating, assisting, and donating to Natives.

Again, like the death of workers or ducks, other forms of loss will be compensated for or superseded in the future. "Liberalism is," Grant writes, "the faith that can understand progress as an extension into the unlimited possibility of the future" (*Lament* 56). This view can be promoted or criticized through literature, and I would like to look at some examples, which can be considered as upward and downward counterfactuals.

Pro and Contra Bituminous Literature

In 1913, only five years after the publication of Mair's *Through the Mackenzie Basin*, Sidney Ells, senior engineer for the Government of Canada's Department of Mines, began his decades' long work to extract oil from the bituminous sands of Northern Alberta. To this end, he undertook surveys, dug test pits and quarries, drilled core samples, and experimented with separation techniques. In addition to writing technical reports, though, he also wrote his memoirs, a collection of short stories and poems entitled *Northland Trails*, and his *Recollections*. In his introduction to this latter volume, Ells writes that in the process of developing bitumen, "Every line and phrase of Kipling's 'If' were...lived" (4). William Dillingham argues that Kipling's worldview, expressed in "If," emphasizes "the role of the hero as one who by an exercise of the will and through self-discipline and inspiration defies the meaninglessness of the universe and the chaos of life by superimposing order—in the form of one's 'work' or 'craft'—upon disorder" (187). Ells positions himself, then, at the beginning of his book, as one of the heroes worthy to be called a "man" by Kipling's speaker. However, that vision of imposing order on self and otherness, which leads to a belief that one possesses the "Earth and everything that's in it" (Kipling) depends on "the fact that the admired spontaneity of freedom is made feasible by the conquering of the spontaneity of nature" (Grant, *Technology and Empire* 31). Gismondi and Davidson, in examining the visual record of representation of bitumen, note,

> While today there are many competing images of the tar sands industry, for much of its history one voice (and its visual conventions or tropes) shaped understandings of the industry's emergence and trials. Most notably, two key patterns cut across the visual record: the application of human ingenuity and scientific and technological expertise to release oil from the chemical bonds of its bitumen form; and corporate and government determination to overcome the physical and economic challenges of opening up the northern frontier to the new industry. (42)

Although the one voice described by Gismondi and Davidson was, no doubt, dominant, a minor thread of doubt was present from the beginning as well. Ells was instrumental in achieving our contemporary freedoms in Canada through both his scientific expertise and his representations of that expertise in discourse, but he was, like Clark, also aware, to some extent, of an order beyond that imposed by human conquest.[14]

In addition to his reference to Kipling, Ells's book begins with a poetic epigraph and concludes with a poëtic epilogue, both of his own creation. The first describes the call of nature to men of various enterprises—"the uplands" to the "shepherd," "the plain" to the "herdsman," "the sea" to the "sailorman," and so on—but, in addition to these vocations,

> now the rock ribbed northland—empire of vale and hill—
> Echoes the thud of bursting charge, the clink of sledge on drill,
> For college don and tenderfoot and seasoned pioneer
> Have turned their faces northward—men of the new frontier! (Ells n.p.)

These lines echo precisely Mair's concluding statement in *Through the Mackenzie Basin*: "The generation is already born, perhaps grown, which will recast a famous journalist's emphatic phrase, and cry, 'Go North!'" (149). Development efforts in the Athabasca region are direct extensions of North American westward expansion in these depictions: an encounter of civilization with wilderness, mediated by the receding horizon of the frontier.

However, Mair's concluding sentiments also contain another side: "modern conditions are breeding methods new and strange, and keen observers profess to discern in our swift development the decay of certain things essential to our welfare" (148). Even though he looks forward with optimism to the coming of "civilization" to the Fort McMurray district, he looks back with nostalgia on a "Nature" already compromised by settler/invader culture: "we had dropped, as it were, from another planet, and would soon, too soon, be treading the flinty city streets, and, divorced from Nature, become once more the bond-slaves of civilization" (146). The young ranger in Kroetsch's *Alberta* (1967) expresses a less ambivalent view of civilization. He is leaving Fort McMurray—because "It's starting to feel like a *city* here"—and moving to Fort Chipewyan. He warns, "If you're writing a book about Alberta don't mention Chipewyan. That *is* Alberta" (280). This binary, between a compromised and compromising civilization impinging on the pristine North and an uninhabitable and backward wilderness in need of taming, remains prominent in representations of Fort McMurray and bitumen extraction today. Viewing bitumen extraction through this lens impedes the possibility of a public committed to building a sustainable community. At the same time, though, denying this conflict is, perhaps, even more common.

Ells, in concluding his *Recollections*, sees no conflict between modernity and wilderness. He writes, imagining his grave marker,

> I asked not stately man-made shaft of stone,
> Within some crowded city of the dead,
> One of a mighty host,—and yet alone,
> While restless feet hurry above my head.

Out on a wind swept ridge then let me lie,
A rugged twisted pine my marker rude,
Where owls' deep call and loons' sad wavering cry,
Alone will break my peaceful solitude.

Yet not alone beneath my tree I'll lie,
For all about me furry things will play,
While stately antlered monarchs wander by,
Friends of the long, long trails of yesterday.

This poem makes no mention of the industrial activity of the first poem and suggests that nature will remain unaffected by bitumen extraction after the speaker's death, allowing friendship with the animal world to continue into the afterlife. Both Mair's and Ells's texts—like Syncrude's sustainability reports and strategic use of bison—thus appear as attempts to balance opposing values: the globalizing forces of liberal, rational, technological progress that promise freedom and equality, and the local, particular, traditional forces that claim virtues of communion and continuity with nature. Or, perhaps more accurately, the forces of progress are shown to pose no threat to the natural world. Gismondi and Davidson claim that, "While engaging, and at times poetic, [Ells's] reports and memoir are largely devoid of attention to nature and ecosystems...Nature was not denied, but circumscribed. At best the muskeg, forest and climate are presented as human trials, obstacles to be conquered...to be overcome by hardy men charged with developing a modern industrial nation" (46). However, that land to be conquered is also loved by these men whose actions threaten its very existence, and Ells's poem can be read as an act of repressing his consciousness of that threat. Nature is an obstacle to be overcome but, once it is overcome, it will still persist unchanged.

Attempts to strike this balance, at least rhetorically, continue in the bituminous sands work being carried out today. The industry works very hard to show that the "thud of bursting charge" and "peaceful solitude" are not mutually exclusive, that the North's "industrial future" will not compromise "certain things essential to our welfare." We see this in Syncrude's representation of the bison.[15] Levant makes this argument as well: "Forests will still grow and critters will still frolic on the land...And even the [land] that is mined will be reclaimed once the oil is pumped out...The first oil sands mine reclamation projects have already been certified. They're gorgeous hiking trails now, with forests and pristine lakes" (*Ethical Oil* 4). Never mind that the single, tiny reclaimed site Levant is referring to was not mined, but only ever used for storing overburden, is nothing like its previous topography, and contains no wetlands. Sacrifice is not necessary to secure the future in this narrative; when the necessary sacrifices are unavoidably apparent, they are justified with reference to that ideal future through the use of upward counterfactuals.

By contrast, Mari-Lou Rowley's poem "In the Tar Sands, Going Down" suggests an alternative to modernity by dwelling in the state of our loss, by remembering our vulnerability, which an imagining of a future (e)utopian community tries to forget. She writes to oil as a "luscious baby," a "perfector of defects" providing "pipelines across continents and so many new / cars, jobs, cans of Dream Whip" (63). As the poem progresses, the speaker has sex while bitumen extraction pollutes the world. While "fondling under leaves" the "sky [is] mortally dazed / under clouds weeping acid" (64). She will continue to satiate her desires while the "boreal blistering" (64) goes on around her, "until," left with "the forestless birds," "the fishless rivers," "the songless, barren face of the earth," "we go down" (66). The "going down" from earlier in the poem, the "knees to ground / head to groin" (64) act of fellatio directed towards escapist orgasmic release from the tragic destruction going on around the copulating humans, in the end includes them (and us) in the exhausted collapse (they and we) were attempting to escape through the pleasures made possible by oil extraction. This downward counterfactual imagining shows the bodily connection with bitumen extraction: we are objectified with the oil we extract.

Though Rowley's poem may be debunked and dismissed as just as ideological as Ells's poems, it is still significant that Rowley's poem critiques one option for life under liberal modernity while Ells's glorify another, and that the one Ells glorifies remains hegemonic. Grant writes, "The vaunted freedom of the individual to choose becomes either the necessity of finding one's role in the public engineering or the necessity of retreating into the privacy of pleasure" (*Lament* 56). Ells found his role in the public engineering, but tried to manage its consequences by imagining an upwardly counterfactual future free from them: we do not see the individual tragedy of the speaker's death, only the pleasant consciousness of his afterlife. Rowley's poem follows our hedonistic retreat into private pleasure through to its grim conclusion. It shows how we are implicated in the public engineering of the oil economy, with the duck deaths, with genocide in Darfur, however far we try to retreat into the privacy of pleasure. Our pleasures, our escapes, our bodies are supported by an oil economy that is inexorably bringing us down at the expense of the beautiful otherness of the boreal forest that is sacrificed to extract bitumen, and the infinity of other otherness in Nigeria, Iraq, Sudan, Libya, Venezuela, Iran, Syria, Russia, and elsewhere.

Robert Kroetsch, shortly before his death in an automobile accident on June 21, 2011, wrote in the first issue of *Eighteen Bridges*, "Story is often about difficult or impossible choices. It challenges our very being. Pure entertainment makes the other guy the goat; it makes us, as mere voyeurs, feel comfortable and privileged. And safe" (19). In the case of bitumen, and modernity more generally, entertainment makes us feel safe by getting us to live, to imagine our relationships to others and other others, in the future. As in Ells's poems, after our deaths everything will be fine. We need more stories about bitumen and less entertainment, less of "the oil sands porn" that Levant says "sells magazines and gets donations" (*Ethical Oil* 10) but also less of the self-satisfied nationalist apologetics that his book provides.

If we can think outside of this binary, it may be possible to reassess bitumen, to think again about the region outside the narrative of progress, to consider "that our response to the world," as Grant had it, "should not most deeply be that of doing...but of wondering or marveling at what is, being amazed or astonished by it...and [to consider] that such a stance...is the only source from which purposes may be manifest for our necessary calculating" (Grant, *Technology and Empire* 35). This would mean, for one thing, reversing the order of Ells's poems; instead of striking bravely forth into the new frontier to establish mastery over it and wrestle value from it, we must imagine first what it would take for places to be ones people could live and die in and then decide how to survive in a way that achieves that goal.

Impossible **Choices**

Fort Mac *and Oil As a "Matter of Concern"*

"CAN YOU IMAGINE if you actually had to *live* here?" When I visited Fort McMurray for the first time in 2008, a long-time resident reported overhearing this statement. That view encapsulates, anecdotally, a flaw in a predominantly binary view of place. The resident wished she had responded with "I can and do."[1] As Mike Gismondi and Debra Davidson write, "In a sense, we have two Fort Macs—one desperately trying to assert a sense of community autonomy and identity, and the other full of people just there to extract paychecks" (92). These views get repeated and entrenched in media representations by people with no connection to the boreal forest of Northern Alberta.[2]

Though some might not want to *live* in Fort McMurray, if a certain wage can be negotiated, they might be willing to come and *work* there. In Marc Prescott's 2007 play *Fort Mac,* Jaypee, Kiki, and Mimi move from Quebec to, as Jaypee says, "faire du *cash*" (20). In Natalie Meisner's work in progress *Boom, Baby,* the main character moves to Fort McMurray from

Nova Scotia to make her pile and go home. Canadian conservative philosopher and critic of technological society, George Grant states that "the retirement of many from the public realm...raises questions about the heart of liberalism: whether the omnipresence of contract in the public realm produces a world so arid that most human beings are unable to inhabit it, except for dashes in followed by dashes out" (*English-Speaking* 12). Fort McMurray has become a place many people dash in to, profit from, and dash out of again.[3]

On the one hand, there are those who look to technological advances within the industry and claim that current environmental issues will be solved by scientific progress. On the other hand, there are those critics of industry and activists who look at the ongoing social and environmental impacts and argue that change is not happening fast enough, or, perhaps, is impossible within the current social mindset. If people don't imagine their futures and their children's futures in a place, then their relationship to it tends to be focused on short-term benefit. The result of framing the debate this way, though, creates the public relations battle that we see currently being played out in the media. Although speakers on both sides muster a wide array of facts, statistics, and data of various kinds, the overwhelming quantity of information makes it impossible to come to any satisfactory conclusion. I suggest, then, following Bruno Latour, that we move away from the model of approaching bitumen primarily with facts and consider it as what he calls a "matter of concern."

Reframing the Debate

In 2010 a coalition of activist environmental groups working under the banner Rethink Alberta, a group ostensibly concerned with adding "facts" to the debate about the so-called oil sands, launched a multimedia ad campaign. The goal was to dissuade international tourists from visiting Alberta by juxtaposing scenes of Alberta's apparently pristine wilderness with ones of the industrial extraction of oil from the bituminous sands around Fort McMurray. As part of this campaign, the group produced billboards, a Facebook page, and a video, which generated responses in newspapers, on television, and online.[4]

Among the dozens of virulent responses posted on YouTube to the video, asing940 writes, "You don't think that this US group has a bias too? Let's see—worst oil disaster EVER just happened in the US. This is an ad campaign, but it isn't designed to trigger interest, it is designed to point the finger somewhere else." In direct response to such anti-bitumen campaigns, the website ethicaloil.org was launched to highlight the differences between "conflict oil" from Venezuela, Libya, the Middle East, etc. and "ethical oil" from Canada. As discussed in the previous chapter, it builds on Ezra Levant's argument in *Ethical Oil: The Case for Canada's Oil Sands* by focusing on the ways in which bitumen extraction creates good jobs and promotes social justice within a regime of environmental responsibility.

The public debate about bitumen occurs within a highly polarized context in which it often seems there is no common ground. This chapter considers how approaching bitumen as what Latour calls a "matter of concern" can interrupt the rhetorical warfare being engaged in by all sides, the binary between "oil sands porn" (Levant, *Ethical Oil* 10) and pro-industry apologetics. In that temporary interruption, an opportunity exists for a shift in perspective.

Latour proposes "an entirely different attitude than the critical one,...a multifarious inquiry launched with the tools of anthropology, philosophy, metaphysics, history, sociology to detect how many participants are gathered in a thing to make it exist and to maintain its existence" ("Critique" 245). Literature also has a role to play. As Travis Mason has noted in a reading of Don McKay's poetry, "That Latour neglects to include poetry, or the arts for that matter, in his project of bringing the sciences into democracy, speaks volumes of the continuing need for sharing ecological consciousness in social spheres that are plural: both public and private, both literary and scientific, both linguistic and kinetic" (Mason par. 26). Though literature too, like any other discourse, can be rhetorical and ideological, it also contains inherent "tensions between content and form, [and]...alertness to the mendacities of language" (Huggan and Tiffin 34), which push against any singular meaning. Thus, it cannot so easily be dismissed as "interested." Of course, propagandist literature exists, but that does not mean literature as such can be reduced to propaganda. As noted in the Introduction, Jean-Luc Nancy claims that literature creates a "circulation [that] goes in all directions at once" (3), and it is in this sense that I use the term, even if actual works of literature fail to live up to this ideal. Literature does not push us towards any of the options in what Robert Kroetsch has called "impossible choices" ("Is This a Real Story?" 19). Literature does not seek to persuade; it is something more than rhetoric or ideology. It allows us to dwell, temporarily, in a present impossibility, the impossibility of choosing between forms of sacrifice: Do we sacrifice, for instance, a particular caribou herd or do we sacrifice small class sizes in elementary education?

In "The Art of Ecological Thinking," Dianne Chisholm argues that "there is an art of ecological thinking which is distinct from ecological science. Science may recognize ecology as a discipline but it does not therefore follow that ecological thinking is properly scientific" (570). There is, she suggests, thinking particular to "literary ecology" and, in the example she addresses—Ellen Meloy's *The Last Cheater's Waltz*—this thinking "deploys affect and sensation in expressive refrains to enact a transvaluation of values" (572). This chapter considers Franco-Manitoban Marc Prescott's play *Fort Mac* as one example of a literary text that creates an opportunity for recognizing the impossible choice presented by bitumen, an opportunity for engaging in ecological thinking, and for deploying affect and sensation to change the prevailing values about bitumen, to see it as a "matter of concern."

Ethicaloil.org, by contrast to the play, claims that we have a clear choice to make between sources of oil: when the choice is framed as one between a place that kills indigenous peoples and a place that employs them, it seems straightforward. However, if it is framed as one between oil-soaked pelicans and oil-soaked ducks, as in the Rethink Alberta billboard, the choice is less obvious. In his book *Ethical Oil*, Levant claims that "the question is not whether we should use oil sands oil instead of some perfect fantasy fuel that hasn't been invented yet...The question is whether we should use oil from the oil sands or oil from the other places in the world that pump it" (7). Putting aside, for the moment, debates about alternative energy sources, and accepting Levant's premise that the world needs oil, if the question is about deciding where that oil comes from, that does not mean there is a simple answer (as Levant's framing of the question implies), and reading bitumen as a matter of concern can help bring the complexity, even the impossibility, of an answer into focus.

Latour defines a matter of concern as "what happens to a matter of fact when you add to it its whole scenography, much like you would do by shifting your attention from the stage to the whole machinery of a theater" (*What Is the Style?* 39). In attempting to follow Latour's suggestion, this chapter shifts between the text of Prescott's play *Fort Mac*, statements made by various participants in debates over bitumen, and theories of discourse and rhetoric. While it is impossible to focus one's attention on all of the factors involved in the discourse surrounding bitumen, just as it would be impossible to focus on the whole machinery of a theatre at one time, this chapter presents one attempt at contextualizing the shortcomings of the current debate about bitumen and pointing towards the (im)possibility of an alternative conversation, a conversation that might consider the ecological view that we need, in Meloy's phrase, to "try to live here [Fort McMurray, Alberta, Canada, Earth] as if there is no other place and it must last forever" (qtd. in Chisholm 586).[5]

"Just the Facts, Please": Objectivity, Control, and "Bullshit"

In Prescott's play, Jaypee, his girlfriend Mimi, and her sister Kiki head to Northern Alberta from Quebec in a dilapidated camper to find the land where it rains jobs. Jaypee is unable to find work because, even though he can fix engines—in his words, he "connais ça des moteurs" (24)[6]—he is not a certified mechanic, and, when he runs afoul of Murdock, a local tough, things quickly become very complicated. Mimi takes a job as an exotic dancer even as she tries to keep her pregnancy a secret from Jaypee. As their need for money approaches a climax, she is unable to make it to work when the truck breaks down. Jaypee becomes addicted to drugs and decides to set up a meth lab to earn enough money to pay off his own drug debts but is unable to buy all the necessary items as the authorities track large purchases of the ingredients for meth: "parce qu'ils font attention à ce que t'achètes" (65).[7] Although Kiki—simple-minded, religious, kind-hearted—tries to help, her best intentions cannot prevent the destructive forces at work. The play presents characters working to change the circumstances they find themselves in, circumstances only partly of their own making and only partly within their control. By contrast, the ongoing rhetorical debate over bitumen suggests, in its emphasis on clear choices, that we do control the situation.

This position has been stated in a, now infamous, open letter published in the *Globe and Mail* on January 9, 2012, where, in the context of the National Energy Board (NEB) opening hearings into Enbridge's proposed six-billion-dollar Northern Gateway Pipeline, which will carry 525,000 barrels of diluted bitumen daily from Bruderheim, Alberta, to Kitimat, British Columbia, Natural Resources Minister Joe Oliver writes, "We know that increasing trade will help ensure the financial security of Canadians and their families." We can exercise control over our future by building this pipeline.[8] However, Oliver goes to say, despite what "we know,"

> there are environmental and other radical groups that would seek to block this opportunity to diversify our trade. Their goal is to stop any major project no matter what the cost to Canadian families in lost jobs and economic growth.
>
> No forestry. No mining. No oil. No gas. No more hydro-electric dams.
>
> These groups threaten to hijack our regulatory system to achieve their radical ideological agenda. They seek to exploit any loophole they can find, stacking public hearings with bodies to ensure that delays kill good projects. They use funding from foreign special interest groups to undermine Canada's national economic interest.

In April 2013, in a speech to the Center for Strategic and International Studies in Washington, Oliver said, "Ultimately this comes down to a choice. The US can choose Canada—a friend, neighbour and ally—as its source of oil imports...Or it can choose to continue to import oil from less friendly, less stable countries with weaker—or perhaps no—environmental standards" (CBC News, "Joe Oliver Slams Scientists"). The issue, as in the Ethical Oil campaign, is a black and white choice between those who would help build future prosperity and those who, for their own political reasons, want to prevent this. Oliver concludes, "Our regulatory system must be fair, independent, consider different viewpoints including those of Aboriginal communities, review the evidence dispassionately and then make an objective determination. It must be based on science and the facts." Nathan Lemphers, oilsands technical and policy analyst for the Pembina Institute, writes in response to Oliver's call for a regulatory system that is fair, independent, and based on science and facts, "We couldn't agree more, and can't help but point out the troubling disconnect between the minister's call for a 'dispassionate' and 'objective' approach and his government's blatant political interference in the process. Remember the BP oil spill in the Gulf of Mexico last summer? *The fact is*, such a disastrous spill could easily happen here, too" (emphasis added). While agreeing with Oliver about the need to base decisions on science, Lemphers then shifts his appeal to the threat of an oil spill, a "fact" that Oliver would surely dispute.

Such exchanges are precisely the problem with the debate currently occurring around bitumen: all sides lay claim to the "objective facts" and defend their positions based on these claims. Change in the debate, let alone government policy or industry actions, will not proceed from more or better facts; we cannot expect some pronouncement from on high that will chart the way forward. Despite what Joe Oliver or Ezra Levant or David Schindler or Andrew Nikiforuk knows, any particular statement of "fact" exists within the context of an ongoing discourse where the various participants are already more or less dogmatic adherents to their positions. Indeed, every attempted truth-claim generates a flurry of opposition activity to debunk it and perpetuate relativist doubt. This doubt, ultimately, serves the status quo. Shifting the debate from facts to concern does not mean abandoning facts or accepting relativism, but putting the facts in a larger context, a context in which the facts appear differently, from which we judge them differently.

For philosopher Harry Frankfurt, "bullshit" is characterized by a whole conversation that is "not germane to the enterprise of describing reality" ("On Bullshit"). Much of the response to Rethink Alberta, and critics of bitumen extraction more generally, has been precisely focused on showing how their facts and statistics do not describe reality. While Rethink Alberta points out, "The Tar Sands are the largest contributor to the growth of greenhouse gas emissions in Canada" (Rethink Alberta), Levant, for instance, replies, "The oil sands combined emit just 5 percent of Canada's total greenhouse gases—less than, for example, the emissions from all of Canada's cattle and pigs" (*Ethical Oil* 6). While Nikiforuk argues, "Bitumen development...will eventually destroy or industrialize a forest the size of Florida" (*Tar Sands* 4), Levant replies, "The oil sands do cover an area the size of Florida. But only 2 per cent of that area will ever be mined" (*Ethical Oil* 4). As far as I can tell, all of these claims are true. However, they point in very different directions. Various individuals and groups would chart a course of action for bitumen based on the "truth," but others dismiss these truths as just so much "bullshit" and chart different courses based on different truths.[9]

Slavoj Žižek argues that in the dominant ideological formation today, "a 'politics of truth' [is] dismissed as totalitarian" (*Defense* 340) because, in the postmodern world, truth is unknowable.[10] However, believing that truth is unknowable serves the interests of the status quo. As William Corlett writes in interpreting Derrida, "In a world where everything is neutralized, the status quo wins" (197). This, perhaps paradoxically, seems to be the result of increasingly partisan, divisive, and extremist rhetoric: increasing the volume of the debate allows extraction to continue apace while a media battle provides a diverting spectacle.[11] While both extremes in the debate I'm exploring make claims to truth, these claims only have purchase for those who already accept them, and can be dismissed as "bullshit" by opponents regardless of whether their positions are more "truthful." Therefore, despite the fact that certain claims about resource extraction have purchase for me, are persuasive to me, I do not believe that my making an argument based on those claims will necessarily have purchase for others. Here I am focused on describing the debate as I see it and suggesting how literature might change that debate by adding to an understanding of bitumen as a matter of concern. Rather than individual facts about bitumen, or even a string of many facts about bitumen, we need to work to see how the facts are all connected. We need to take a lyric perspective, as I discuss in Chapter 1, to consider bitumen as a matter of concern.

In developing the concept of matters of concern, Latour quotes a Republican strategist from a *New York Times* piece regarding the debate over climate change: "Should the public come to believe that the scientific issues are settled," the Republican writes, "their views about global warming will change accordingly. Therefore, you need to continue to make the lack of scientific certainty a primary issue" ("Critique" 226). This leads Latour to worry:

> I myself have spent some time in the past trying to show "the lack of scientific certainty" inherent in the construction of facts. I too made it a "primary issue."...Was I foolishly mistaken? Have things changed so fast? In which case the danger would no longer be coming from an excessive confidence in ideological arguments posturing as matters of fact—as we have learned to combat so efficiently in the past—but from an excessive distrust of good matters of fact disguised as bad ideological biases! ("Critique" 227)

Things are even more complicated than Latour shows, as, on the one hand, this strategy of creating distrust of facts is employed for any negative science, but, on the other, scientific certainty is asserted about the potential for continuous improvements in extraction processes. In *Ethical Oil*, Levant provides examples of both of these strategies: he writes, for the first, "It's true, there is oil seeping into the rivers north of Fort McMurray and sometimes the air smells like sulphur and the water is bitter. And that's how it's been for millennia" (4), and, for the second, "Oil sands technology continues to improve—to produce one barrel of oil sands oil takes 38 percent less [greenhouse gas] emissions now than it did in 1990" (6). In this way, any doubts about the negatives that remain after the industry spin machine is finished with them are deferred into the future.

Mustn't we, then, consider every claim made about the development to be more or less bullshit? How can we know what the facts are, whether we like what they say or not? And, further, given that the critical work of people like Latour has been to cast doubt on the possibility of scientific consensus ever being established, *and* that such techniques have been appropriated by "the other side" as Latour argues, *and* that the promise of bitumen development has always been in future technological advances that would make it economically and environmentally viable, can there ever be anything said about it that is not, or, at least, cannot be dismissed by parties with opposing interests as, bullshit? For example, William Marsden quotes Rod Love, Ralph Klein's former chief of staff as stating, "the Conservatives regarded most of the professors who opposed the government as 'flakes.'" Marsden quotes Love as saying, "[people like David Schindler] weren't scientists, they were politicians...We would say 'This is not science. This is a political agenda not an environmental agenda['] and they didn't like it" (105). As Latour points out, facts have come to serve the role of "shut[ing] the dissenters' voices down" ("Style" 39).

With matters of concern, though,

> It is the same world, and yet, everything looks different. Matters of fact were indisputable, obstinate, simply there; matters of concern are disputable, and their obstinacy seems to be of an entirely different sort: they move, they carry you away, and, yet, they too matter. The amazing thing with matters of fact was that, although they were material, they did not matter a bit, even though they were immediately used to enter into some sort of polemic. (Latour, "Style" 39)

The NEB hearings into Northern Gateway should, ideally, provide one opportunity for a conversation that moves beyond the polemic of matters of fact, but it is also in danger of being hijacked by paternalistic claims to transparent access to universal Truth by government and industry, and, indeed, by environmentalists as well.[12] If we accept that the starting point for debate is "science and the facts," as Oliver asserts and Lemphers accepts, then the terms of that debate are dangerously narrowed.

In "Canada and Postcolonialism: Questions, Inventories, Futures," Diana Brydon argues that postcolonialism depends on the recognition that

> all truths are complicated and contingent; while there may be many truths rather than a single Truth, that does not absolve an individual or a community from distinguishing among them nor from establishing priorities, nor indeed from seeking consensus through discussion and compromise; and that Eurocentric forms of truth have masqueraded as the universal under a hijacked form of humanism; but that it remains

> necessary to search for ways to create a renewed definition of the human, beyond the commodification of identity under capitalism. (73)

When Oliver writes that the regulatory process must "consider different viewpoints including those of Aboriginal communities," but then asserts that, "It must be based on science and the facts," while clearly implying that the pipeline should be built, he repeats a colonialist disavowal of indigenous knowledge and concerns. A conversation about the various complicated and contingent truths at play in bitumen development and what the communal priorities should be is the one that we are struggling to have in Alberta, in Canada, and internationally.

If we consider bitumen, and the ecosystem of which it is a part, a "matter of concern," rather than a substance about which we just need to establish scientific facts, then the process of seeking consensus through discussion and compromise, of establishing priorities, becomes meaningful—more difficult, yes—but also more democratic: a process that does not privilege Eurocentric models of knowledge over every other includes more truths about bitumen and the boreal forest. In "Why Has Critique Run Out of Steam? From Matters of Fact to Matters of Concern," Latour claims,

> Archimedes spoke for a whole tradition when he exclaimed: "Give me one fixed point and I will move the Earth," but am I not speaking for another, much less prestigious but maybe as respectable tradition, if I exclaim in turn "Give me one matter of concern and I will show you the whole earth and heavens that have to be gathered to hold it firmly in place?" For me it makes no sense to reserve the realist vocabulary for the first one only. The critic is not the one who debunks, but the one who assembles. (246)

A critique of assembly is desperately needed in the debate over bitumen development. While various participants struggle to establish fixed points by which they can move the world's understanding of bitumen and debunk opponents' claims, this strategy has not succeeded in changing the status quo. As water ecologist David Schindler and others have pointed out, there has been a tremendous lack of environmental monitoring in the region, and what monitoring has been done was largely been done by industry. The implementation of the Canada-Alberta Joint Oil Sands Monitoring Program, which was announced in 2011 but did not start its three-year phased implementation until 2012, remains primarily industry-funded, though the monitoring results are now publically available and the system will undergo periodic peer review.[13] Further, the critique of debunking is so firmly entrenched that it is very difficult to know which points are truly fixed. Thus, when 99kokanee writes in response to Rethink Alberta, "do you deny *the fact* that the oil is naturally seeping into the river causing the toxins. do you deny *the fact* that water sampling and testing has shown no increase in toxins in the last 50 years," (emphasis added) it is impossible, at least for most ordinary citizens, to definitively dispute this claim. Despite Erin Kelly and David Schindler's research, published in the prestigious *Proceedings of the National Academy of Sciences,* which shows that "not all toxic metals in the river are from natural sources" (Brooymans, "Oilsands Boosts Toxic Metals" B4), and federal and provincial government promises to increase monitoring, the conflicting narratives about the impacts of bitumen extraction have so far preserved the status quo, which privileges development over conservation.[14] Schindler's facts have not gained more traction than those of industry or government. If it is a fixed point, a fact, that industry is depositing deleterious substances into the Athabasca River, in contravention of the Fisheries Act, this has not changed industry practices or reframed the debate. Indeed, the federal Conservative government has, with its omnibus bills C-38 and C-45 made changes to the Fisheries Act and the Navigable Waters Act that weaken environmental protection. Attempts to work within the existing legislative context to show the dangers of bitumen extraction have led to changes to that context to exclude the concern. While Schindler does receive media attention, and his views are taken seriously by some, he remains one voice among many. His warnings about current pollution and future dangers have not punctured industry's balloon, inflated by jobs, taxes, funding for arts and culture, and promises of continued prosperity and improved environmental performance. The facts that Schindler has established have not led to widespread concern such that industry has had to fundamentally change the way it operates.[15]

If the goal is to change the debate, then a strategy that not only searches for more or better facts but also assembles the various voices that enable the status quo as a way of trying to see how they operate together is worth exploring. TheMountainDude8, who writes in the same vein as 99kokanee, states, "The oil companies in Alberta are smart enough to do something about it. They are investing millions in research and working with local researche[r]s at the Universities to find better ways of extracting the oil, minimizing the footprint, and sequestering CO_2 in old depleted wells." An opponent may well respond that the open pit mining by which much bitumen is still being extracted does not use wells, that in situ extraction still requires huge amounts of water and natural gas and fragments habitat, that carbon capture and sequestration is not being done on a commercial scale anywhere, that serious doubts about its feasibility remain (doubts strengthened by the 2013 cancellation of high-profile carbon capture and storage [CCS] projects co-funded by the Alberta government [Blackwell]), and that the technology is generally unsuitable for use in the bitumen region.[16] My main interest in 99kokanee's statement, however, is the rhetorical effect of the ambiguous pronoun "it." He or she does not say explicitly what the oil companies are smart enough to do something about. The sentence that follows suggests that "it" is the environmental damage wrought by bitumen extraction, but we might also read "it" as the *perception* that bitumen extraction is environmentally destructive: the oil companies are smart enough to do something about the negative perceptions associated with their industry. A cynic might suggest that all of the efforts 99kokanee lists are targeted toward this perception as much as, or more than, the destruction. As Schindler himself put it, "I guess what really rankles me is we have this foolish propaganda going on to make people think everything is OK and therefore support the tarsands," but "If people know what's really going on and they still support the tarsands, well, I can grit my teeth, but that's democracy. What's going on now is not democracy" (qtd. in McLean and Brooymans A2). The challenge for a position like Schindler's, though, is dealing with the counterrhetorical pincer move of dismissing such a claim to know "what's really going on" as totalitarian on the one hand and falling back on industry's "objective" claims to improvement on the other.[17] We need to supplement efforts to describe what's "really going on."

Propaganda and an Attentive, Affective Alternative

We can see the type of propaganda Schindler criticizes at work in a piece that Dave Collyer, president of the Canadian Association of Petroleum Producers (CAPP), wrote for the *Edmonton Journal* on August 11, 2011 titled "Oil, Gas Industry Is Making Environmental Progress." He writes, "The Canadian oil and gas industry...is focused on the three 'E's'—energy security and reliability, economic growth and environmental performance"; this would seem to indicate that these can all be focused on at the same time, that a balance can be achieved.[18] Collyer articulates precisely a "need" for "balanced solutions—responsibly developing Alberta's valuable resources within a reasonable regulatory framework that recognizes economic benefits and the need for practical environmental protection." Further, "In addition to focusing on environmental performance improvement, we must continue to improve our engagement and communication with customers and the Canadian public," but "Unfortunately, we also understand responsible environmental performance and objective communications won't satisfy the activists who oppose oilsands development." This type of ad hominem attack has been extended in the federal government's intervention in the debate over the proposed Northern Gateway Project. Joe Oliver's attack on "radical" environmentalists and "jet-setting celebrities" who seek to "undermine Canada's national economic interest" attempts to marginalize anyone with doubts about the project.[19] Oliver has also stated in interview, though, that, "There have been very few pipelines indeed that have ever been rejected by the National Energy Board. I think there have only been two out of tens of thousands" (O'Neil, "Feds" A5). Andrew Nikiforuk, writing on the other side of this "war of words," quotes unconventional oil expert Maurice Dusseault: "Giving the [government] agency responsible for production and royalties [the Energy Resources Conservation Board (ERCB)] the mandate to also enforce [environmental] regulations leads to a difficult internal conflict of interest. The result is usually a clashing of goals. Most commonly it is the enforcement of environmental regulations that suffers" (*Tar Sands* 30).[20] Collyer's next claim is that, "because all forms of development impact the environment, industry's role is to reduce its environmental footprint to the extent reasonable and practical within a responsible regulatory framework." This begs all kinds of questions: What is "reasonable and practical"? Who decides? Is there currently a responsible regulatory framework? Collyer seems to imply that such a framework is in place and is working well, while opponents of development argue that the system is broken and toothless.

While Collyer and Oliver are targeting groups such as Greenpeace and Corporate Ethics International, the coalition of organizations behind the Rethink Alberta campaign encouraging tourists not to visit Alberta, and individuals like Nikiforuk, one can assume that they would also take issue with Prescott's *Fort Mac*, which suggests, according to one review, that the boomtown lifestyle "will steal your soul" (National Arts Centre). However, Prescott's play presents a series of events that cannot simply be refuted with a set of statistics. We might think of these events within what Chisholm calls "the art of ecological thinking" or what Adam Dickinson has theorized as "lyric ethics," which is understood as "a kind of attention that is not reducible to linguistic code or description, a form of listening, perhaps, that might serve to hear the imperative of the other, human and nonhuman" (48).[21] In short, literature pays attention to ecology, recognizes it as a "matter of concern," in ways that science and politics do not.

Commissioned by Daniel Cournoyer for Edmonton's French-language L'UniThéâtre at the height of the 2008 boom, *Fort Mac* went on to run at the National Arts Centre in Ottawa. Reviewing the play, Alexandre Gauthier wrote that the interest of the piece is in the characters, as they are thrown into a world governed by money:

> Et c'est là l'intérêt de la pièce. Fort Mac n'est pas qu'une fable écologique qui s'emploie à dénoncer l'exploitation des sables bitumineux. Ce n'est pas non plus qu'une pièce politique, revendicatrice d'une position anti-capitaliste claire et précise. Il s'agit tout simplement d'une pièce sur l'humain—l'homme, la femme—qui, vouant un culte à l'argent et à la prospérité, entrent dans un tourbillon sans jamais pouvoir en ressortir. Les personnages dévoilent peu à peu, directement au public, des fragments de leur humanité, humanité qui s'effrite tout au long de la pièce.[22] (Gauthier)

It is the humanity of the characters that makes it impossible to discount the play as a piece of rhetorical propaganda. As Cournoyer describes them, "to forget the past and reinvent themselves....Crawling out of your old self: That's a story everyone understands" (qtd. in Nicholls). The way that the play catalogues the failure of the characters to achieve this re-creation in Fort McMurray, and the failure of Fort McMurray to enable this re-creation, casts doubts on Joe Oliver's certainty in bitumen development and pipeline building to generate a better future for all. Some people remain marginalized.

As previously mentioned, Fort McMurray is often seen as a place people dash in to, profit from, and dash out of again, despite the community-building work of many inhabitants and non-profit organizations. In the play, this movement seems, indeed, to be a necessary survival strategy. As Gauthier notes, Kiki is the only character to believe in something other than money—"Kiki, en vierge Marie des temps modernes, prie, en vain. Elle est bien la seule à croire en autre chose qu'en l'argent"[23]—and, because no one else shares this belief, she dies in the end.[24]

Kiki begins the play "sur un pont" (9) (on a bridge) contemplating suicide, but decides not to kill herself because she feels her death will not have any meaning: "Si je meurs maintenant ma mort aurait pas de sens" (14).[25] Instead, she attempts to build a life of mutual benefit for herself and those around her, trying to help in any way, however ineffectual, whenever she can. She says, after finding work at Tim Hortons, that making people happy is her reason for being: "J'ai trouvé ma raison d'être. Je vas render les gens heureux" (35).[26] Ultimately, even though she offers herself as a sacrifice to pay the debts of those she cares for, it seems unlikely that she has made anyone happy. Her intentions are thwarted. Gauthier describes how Kiki represents the dangers of human greed: "Kiki est la seule à reconnaître les dangers de l'exploitation des sables bitumineux, invoquant souvent Dame Nature. Mais elle représente surtout les dangers humains liés à l'appât du gain."[27] While one might dismiss the invocation of Mother Nature as an appeal to a sentimental construct, it is important, in my view, if we are to establish bitumen as a matter of concern, that we not categorically dismiss such appeals: sentiment may provide access to a truth, one that is marginalized within Eurocentric rationalist discourse, but one that may usefully be redeployed.[28] The selfishness of others leads to Kiki's tragic death, and if her death is more meaningful at the end of the play than it would have been at the beginning, it will be because of the audience's recognition of the dangers to human and non-human nature that come with greed, that come with all people acting only in their own self-interest. This ending is meant to evoke an emotional response from the audience, certainly, and that response enables a change in perspective, a change to a perspective that recognizes relations that exceed the contractual. Daniel Coleman has argued that "reading a thing that is truly admirable...reminds us what it is like to be open and undefended." He adds, "We need this reminder...because it opens us to the Other" (37). Being open to otherness, I would suggest, is a requisite step in moving from matters of fact, and the polemical rhetoric and "bullshit" that they generate, to matters of concern, where the beauty of otherness is recognized as good for its own sake. Admirable texts may come from many realms, but there is something about how we read fiction, and experience art, that can facilitate this openness. Recognition of beautiful otherness does not absolve us of the hard work of establishing priorities that Brydon outlines; indeed, it shows us how important that work is when we recognize the other as beautiful and at risk.

If, as Ian Angus, argues, "the concept of community refers," on one hand, "to the human community within which the individual lives and works" but also, on the other, to "the natural environment that surrounds and sustains human life" (*Emergent* 84), ecological and communal consequences arise when it becomes impossible to imagine both living and working in a single location, impossible to imagine living with the consequences of your work. Kiki is beaten to death while prostituting herself in a work camp in an attempt to earn enough money to pay off Jaypee's debts. Her death illustrates a breakdown of the sustaining human and non-human community in Fort McMurray. In the first "Intermède," a series of soliloquies that separate the "Scènes," Kiki asserts an etymological connection between "nature" and "naissance" before arguing, "Si Dame Nature voyait ce qui se passait ici à *Fort McMurray*, elle pleurerait son océan de larmes. Mais là que j'y pense...Elle n'a pas besoin de le voir—elle le sent, elle le sait. Mon Dieu...Elle doit le sentir" (16).[29] The breakdown of that sustaining surround leads to her death. Despite the help of the well-intentioned Maurice, the Franco-Albertan from Plamondon who has come to Fort McMurray to escape the small-town pity and gossip that followed his wife leaving him, Kiki is not able to create community in Fort McMurray, despite sacrificing herself in the attempt.

The Rhetoric of Bitumen Extraction and Its Limits

The representation of Kiki's sacrifice pushes beyond the rhetorical battle, the "war of words" (A15), as CAPP president Dave Collyer has characterized it, in which, while there may be some fixed point somewhere that the committed citizen can uncover if given the resources, for the average observer it all comes to resemble so much "bullshit." That is, all participants seem to be concerned with is persuasion "unconnected to a concern with the truth"; the whole conversation is "not germane to the enterprise of describing reality" (Frankfurt). Regardless of the intentions of the sender of the message, which are more or less unknowable, this is how it appears to the receiver. Revealingly, numerous posters declare that the Rethink Alberta video is "bullshit." Gismondi and Davidson claim of the Rethink Alberta campaign, "The success these organizations have had in creating these discursive openings in the international political arena should not be downplayed; the fact that an industrial enterprise on a remote frontier has gained international stature, inspiring direct actions all over the globe, is nothing short of remarkable" (179). However, "the very selection of certain frames by definition diverts attention away from others, and those others may be quite significant to instigating a post-oil transition to a low carbon economy. Most notably, these activist frames do not resonate with the frames that have been embraced by that population we consider to be of greatest interest—concerned Albertan citizens" (179). While it is impossible to know if the YouTube posters are Albertans or not, it is clear that some accept and others reject the frame established by Rethink Alberta. In rejecting this framing, many posters instruct others to "get educated" about bitumen development and turn to "facts," such as KakeC13, who writes, "Do some research and you will find out that the Province counter acts everything, they take down trees, we plant them." The ambiguous pronouns *they* and *we* create an interesting opposition, since the oil companies are responsible both for taking down the trees and planting new ones. KakeC13's us vs. them dichotomy makes industry both us and them, appropriately both self and other since all Albertans, Canadians, North Americans are implicated in the environmental degradation required to extract bitumen and the costs of its regeneration. Frankfurt argues, "Bullshit is unavoidable whenever circumstances require someone to talk without knowing what he is talking about. Thus the production of bullshit is stimulated whenever a person's obligations or opportunities to speak about some topic are more excessive than his knowledge of the facts that are relevant to that topic." I don't mean to demonize the YouTube posters, or simply expose their ignorance. Talking about bitumen, perhaps talking about the environment in any capacity, requires talking without knowing what we are talking about (we are not omniscient and this is one reason perpetuating doubt proves such an effective strategy). There is, certainly, good reason to take issue with the manipulative way that "facts" are used in both the Rethink Alberta video and the Ethical Oil campaign, as I have tried to suggest above. My point, rather, is that despite the proliferation of statistics marshalled by all sides in the debate, there is a lack of fixed points about the consequences of bitumen development, and this is the situation that we need to come to terms with, the circumstance we need to learn to live within. As Allan Stoekl argues,

> Just as cost and the origin of all those costs is unknowable, ungraspable, we have a motive to keep them unknowable...The unknowable, ungraspable urge to spend, to consume, to burn—by definition irrational, given all the consequences of the act, themselves (or their costs) ultimately unknowable—along with the willful desire, kept hidden no doubt, not to know. We know enough to want not to know...Ultimately the future of "unconventional oil" may boil down not to precise calculations through which it can be known and controlled—though all that is necessary—but to its role as an agent in which our very subjectivities are both constituted and called into question. ("Unconventional Oil" 40)

We do not and cannot know what the results of bitumen development will be. In such a situation, rather than trying to exclude everything that does not provide an indisputable matter of fact from the conversation, as Oliver claims we must, we can "follow the poets in their quest for reality" (Latour, "Style" 23), which I take to mean that poetic (or literary, or more broadly artistic) descriptions of reality get beyond matters of fact by facilitating meta-cognitive reflection on the ways in which our subjectivities are constituted and by exceeding the intentions of the "author." Matters of concern are a gift that can break through our desire not to know and help us think in new ways.

Tell It Like It Is: Enthymemes, Identification, and Unknowns

Bitumen production occurs because a majority of citizens want to believe it can satisfy our appetites without any permanent sacrifices, that reclamation can return the land to its pre-development state, and that pollution is compensated for by jobs, taxes, and increasing prosperity. Gismondi and Davidson write of this situation, "Whether Albertans privilege their 'right' to produce the raw materials located within their jurisdiction, or instead privilege the costs borne by others as a result of that development, defines a primary mechanism through which tar sands development is deemed legitimate or not" (180). However, "If those Others (First Nations, future generations of Albertans, victims of global warming in other countries, or other species) cannot be clearly identified as members of a shared community of citizens, then consideration for them will lie outside the realms of obligation, requiring expressions of virtue" (180). Further, "in the area of environmental politics, negative impacts can fairly readily be disregarded or explained away because of the limited ability to perceive and assess impacts without specialized training and equipment" (180). In addition to these factors, the belief in bitumen production's legitimacy is based on a dominant cultural enthymeme—a syllogism in which some premises remain unspoken and that the audience must supply. Aristotle saw the enthymeme as the most important rhetorical form (78).

When some premises are suppressed or assumed by both sides in a debate, appeals to *logos* reach a limit. As Aristotle notes, in situations "where precision is impossible and two views can be maintained," the logical appeal must give way to the ethical—the perceived character of the speaker—as we "sooner believe reasonable men [*sic*]" (75). This helps explain why responses to criticism about bitumen development have focused so heavily on public relations, and perhaps suggests the opportunity that current debates over Northern Gateway and Keystone XL offer.

The Alberta government, for example, famously took out an advertisement in New York's Times Square, which stated that, "a good neighbour lends you a cup of sugar. A great neighbour provides you with 1.4 million barrels of oil per day" (Government of Alberta, Times Square Advertisement). The tagline for the campaign was "Tell it like it is." This line asks the reader to fill in the premises and, thereby, co-operate in creating its persuasive effect. Reading this line with one set of premises would give, "Tell bitumen's story accurately (not as environmentalists tell it): it is being done properly by a neighbour of strong moral character and is beneficial for all." The ambiguity of the pronoun referent—the "it" of the tagline—creates space for counterdiscursive responses as well, but the effectiveness of such responses depends on recognition of what Žižek (adding to Donald Rumsfeld's infamous philosophizing in 2003 about known knows, known unknowns, and unknown unknowns) has called "unknown knowns" (*Defense* 457). This is, Žižek writes, "precisely the Freudian unconscious," the "disavowed beliefs and suppositions we are not aware of adhering to ourselves...[which] in the case of ecology...prevent us from really believing in the possibility of a disaster" (457). The ambiguity of the pronoun referents—the double "it" of the tagline—creates space for counterdiscursive responses, such as author and environmentalist Chris Turner's "Paradigm Shift," a letter written to the next premier of Alberta (before the 2012 election of Alison Redford), which claims that the reality is, "The tarsands will be a curse if not managed properly. It can also be a great gift to Canada and Alberta" (32). Turner's appeal, though, remains within the logic of management: humans can control the development of this resource so that the ecological curse can become an economic gift. To quote Žižek again: "therein also resides the limitation of the 'ethical committees' which pop up everywhere to counteract the dangers of unbridled, scientific-technological development: with all their good intentions, ethical considerations, and so forth; they ignore the more basic 'systemic' violence" (458). The government's approach tends to be

FIGURE 9: "A great neighbour" Government of Alberta Times Square Ad, 2010.
[Government of Alberta]

effective, because, in asking readers to fill in the blanks, it appeals to the cultural dominant and re-establishes the status quo, a dominant which even Turner's attempted rereading does not escape. When the "unknown knowns" combine with the "unknown unknowns," Žižek writes, "The situation is like that of the blind spot in our visual field: we do not see the gap, the picture appears continuous" (457). Such is the rhetorical power of the enthymeme: we fill in the gap in the logic automatically. Even when, as with Turner's letter, such a trope is read catachrestically, it does not make us aware of the gap: it simply fills it in differently. The unknown knowns, our disavowed knowledge about sustainable extraction of a non-renewable resource (we know it is impossible), are combined with the unknown unknowns about long-term ecological consequences of such extraction (we cannot know what these will be, and do not even know, cannot predict, what it is that we do not know about these consequences). Literature, by contrast, though it calls on readers to create the meaning of texts in similar ways, can interrupt the operation of that status quo, at least temporarily, by making readers reflect on the unknowns.

As noted in the Introduction, Kenneth Burke describes the "characteristic invitation to rhetoric" this way: when partners "collaborate in an enterprise to which they contribute different kinds of services and from which they derive different amounts and kinds of profit, who is to say, once and for all, just where 'cooperation' ends and one partner's 'exploitation' of the other begins?" (25). Alternatives to the status quo emerge when we recognize the beliefs whose disavowal allows us to represent our exploitation of nature as co-operation. However, as Burke notes, "a 'good' rhetoric neglected by the press obviously cannot be so 'communicative' as a poor rhetoric backed nation-wide by headlines. And often we must think of rhetoric not in terms of some one particular address, but as a general *body of identifications* that owe their convincingness much more to trivial repetition and dull daily reënforcement than to exceptional rhetorical skill" (25). Gismondi and Davidson point to the depth of citizen testimony at the Oil Sands Consultation hearings and how the hearings "backfired" as a government "exercise in symbolic legitimacy maintenance" (177). However, while the process may have failed, in the eyes of those who testified, to legitimize government actions, for those citizens outside of the process the knowledge that a process of consultation was occurring may have served to create legitimacy, especially when combined with the repetition of the narrative of government as responsible regulator and industry as technologically innovative.

The heavy-handed intervention of the federal government in the Northern Gateway debate, though, may enable a broader recognition of the disavowals required for the body of identifications built into the dominant rhetoric. The failure of Oliver's rhetoric—to appeal to the ethos of those who do not consider themselves "radical environmentalists" and yet have doubts about running a pipeline across hundreds of rivers and streams or tanker traffic off the treacherous BC coast where a spill would threaten one of the world's largest intact temperate rainforests—in this particular instance, could serve to make people more aware of the disavowals required to identify with the general rhetoric of sustainable development. Campaigns such as Rethink Alberta and Ethical Oil, by contrast, in their blatant appeals to emotion, are so polarizing that they only reinforce the positions of those who already hold them. They fill in the premises of the syllogism with a predetermined script: bitumen development is either a disaster or an ethical good. There is no middle ground. The success of government and industry positions may, in part, be attributed to their ability to present themselves as occupying such a middle ground. Gismondi and Davidson describe the Progressive Conservative government's management of the discourse this way, "PC narratives are strikingly consistent, including characterization of industry as responsible citizen, that of environmentalists as emotional and irrational, and self-characterization as reasoned, calculated leaders" (134). They go on: "The Premier [Stelmach] has gone so far as to flip the environment/development relationship on its head, suggesting that tar sands development is necessary for generating revenues that will drive the greening of our economy. Throughout this narrative, particular frames emerge, including prescription to technological optimism, and invoking of Western cultural identity that espouses individualism and determination, all the while suggesting that placing controls on growth offers the greater threat" (135). The idea that bitumen extraction will pay for the greening of the economy is an increasingly popular position among "progressives" (see Chapter 1). Since being elected in May 2015, the New Democratic government has taken pains to avoid being characterized as anti-development and, in so doing, has been recycling some of the rhetoric of the previous government.

Unsustainable Oil does not argue that literature seeks such a middle ground, but it can present an alternative view of reality, one that people with various views might recognize as having a place in the conversation. This recognition can help to counter the general lack of fixed points. Indeed, among the specifications Latour offers for a project based in matters of concern is that they "have to be *liked*" ("Style" 47). This suggests that, for one, in making decisions about bitumen development or pipeline building we should listen carefully to the people who live in, and like, the places affected. This may be difficult for Oliver who seems to think the land holding bitumen deposits, "is uninhabitable...uh...by human beings. So, you know, no community is being disrupted" (Paris). It is shocking that the colonialist view of *terra nullius* could be maintained by anyone, let alone a contemporary government minister, and not even to describe the past but to describe the present situation in Northern Alberta. This may offer an example where critical debunking still has a role to play. The tools of a hermeneutics of affirmation, though, can be brought to bear to give a sense of why we care that some thing is made to exist or should be allowed to maintain its existence. The intention is *not* to be objective, *not* to "let the facts speak for themselves" as Collyer suggests he is doing. By considering *Fort Mac* we can see how moving out of the realm of facts temporarily, and into that of the values that language can only intimate, might help us change our minds about what our relationship to bitumen should be.

Kiki explains her difficulty in understanding how men can dominate nature when recognition of the beautiful otherness of nature makes us human:

> J'ai jamais compris le besoin qu'ont les hommes de vouloir détruire ce qui est beau. Je comprends juste pas. Peut-être ils voient pas la beauté qui les entoure—mais je refuse de croire que ces hommes peuvent pas voir la beauté parce que c'est ça ce qui nous rend humain. Pis je crois pas que ces hommes sont pas humains. Ils se sentent peut-être à part—détachés ou supérieurs à la nature. La nature est seulement une force à dompter—à dominer—à exploiter.[30] (67)

Jaypee shows himself to be precisely such an exploiter as he encourages Mimi to take a job as an exotic dancer, convinces her to prostitute herself to Murdock (who never appears on stage), and, as soon as Mimi has left to try to save Jaypee from Murdock, he tries to force himself on Kiki. When she asks him what he wants from her—"Qu'est-ce que tu veux de moi?—he states, "Je veux ton innocence. Ta lumière. Ta beauté. Ta bonté" (98).[31] However, when she doesn't resist, saying, "Si tu me veux, prends-moi," he is unable to rape her, saying under his breath, "(*Fuck*...Je peux pas fourrer la Sainte Vierge, estie!)" (98).[32] Those like Jaypee—and the more malevolent, and successful, Murdock—who see all forms of otherness as potential to dominate and profit from, are certain in their own knowledge regardless of any evidence to the contrary. Throughout the play, whenever Jaypee requires support for some claim he is making he says, "J'ai raison, j'ai pas raison?" and then he will wait for Mimi or Kiki to reply, "T'as raison," before concluding with "J'ai raison, çartain" (22).[33] As the play moves on, however, and he gets further in debt and more drug-addled, the supportive response of others is no longer forthcoming, but this does not cause him to question his certainty. He simply waits a moment before declaring, "Çartain que j'ai raison" (97). Jaypee's certainty is easy to dismiss as flawed. There is very little appeal in his ethos to come to his defence when the limits of his reason are reached.

When Patrick Moore, co-founder of Greenpeace, however, campaigns for CAPP on behalf of Northern Gateway, describing post-mining restoration as "the best reclamation I've seen," his ethos provides more support. In an op-ed, Moore writes, "the world needs oil now and we'll need it for the foreseeable future—so it matters greatly where that oil comes from. If any oil is to be labeled 'dirty,' shouldn't it be the oil coming from dictatorial regimes...?" (A19). Despite such efforts to show the superiority of bitumen, both *logos* and *ethos* reach their limits in moments when particular horrors cannot be repressed, dismissed, or marginalized. It does matter greatly where oil comes from; we may even decide that bitumen is the best source; but the extraction of bitumen does require moments of sacrifice that cannot be compensated.[34] The conclusion to Prescott's play offers one such moment, as do the deaths of the ducks that Rethink Alberta attempted to capitalize on. Such moments do not prescribe a course of action or attempt to chart a way forward; their value is in the way they raise the question of what should be done, of whether anything will be done in response to such tragedy. Continuing with the status quo is a possibility, probably the most likely possibility, but Kiki's death makes us either choose the status quo again or choose something else.

Choosing to Pay Attention to Ambiguity

In the final Intermède, Maurice describes the abuse of nature by humans: "Dame nature est fatiguée, aussi. Elle s'est offerte a nous avec toutes ses bontés pis on la remercie en profitant d'elle. On prend avantage d'elle, on l'abuse...On la viole" (114).[35] We can read Kiki as a representative of nature. Maurice stands on the stage waiting for recognition of her sacrifice—"pour marquer son depart, sa souffrance, son sacrifice" (114)[36]—but a sign is not forthcoming from nature and he must mark her departure himself: before the stage goes dark, "Il lance des pétales de roses en bas du pont" (115).[37] It may be a literary-critical commonplace to point out that the ambiguity of this conclusion leaves the possible truths of the play open to the interpretations of the audience members. Nevertheless, that openness is important and the type of reading it requires—debate, discussion, and compromise—to make whatever sense it will make needs to be extended to the conversation about bitumen in general. Such reading works against the unconscious gap filling of enthymemic rhetoric. The debate such reading could prompt, one that considers all of the competing truths before establishing a set of priorities through a process of discussion and compromise, would characterize an understanding of bitumen as a matter of concern.

Today, however, instead of engaging in the type of collaborative debate required for textual analysis, battle lines are being drawn. The "good" rhetoric of collaborative debate does not have the same access to the "technical means of communication" (Burke 25) as the "bad" rhetoric of unconscious gap filling. As mentioned, the federal government has taken an aggressive line against foreign influence in the anti-pipeline movement and has taken steps in its 2012 budget implementation bill to restrict the opportunities for charities to undertake political activities, attempting, thereby, to limit their access further. The January 10, 2012 editorial in the *Edmonton Journal*, "Foreign Influence Foolish Target," questions this tactic:

> By using assaults on the legitimacy and credibility of opponents instead of engaging with adult politeness, the government has given a clear signal which way it is leaning before the independent panel gives the first formal testimony a respectful hearing. Even worse, it has undermined its

credibility by being inconsistent, if not outright hypocritical, on the subject of foreign participation in the debate. After all, the foreign money being poured into the anti-pipeline fight is pocket change compared to the billions that French, American, Norwegian, and Chinese business people have invested in the oilsands. (A12)[38]

The editorial concludes by stating, "name-calling isn't how you win this argument; on the contrary it's a surefire way of getting your opponents even more committed to doing battle" (A12). Indeed, rather than helping to win, name-calling might help to create the conditions for losing the argument if it helps to undermine the pro-bitumen side's ethos. If we "sooner believe reasonable men," continued evidence that defenders of the status quo are not more reasonable than their interlocutors might create the conditions for change. As Žižek argues, "Words are never 'only words'; they matter because they define the contours of what we can do" (*First as Tragedy* 109). Literature is one place where the definitions of words can be changed, where the contours of what we can do can be changed. In *Fort Mac,* Prescott's attention to the multiple sacrifices required by bitumen extraction, and connection of human and non-human sacrifice in the character of Kiki, provide one opportunity to change the debate from one that starts and ends with scientific facts to something broader. It is easier in a play to argue at a remove from immediate factuality and necessity, and that remove enables a type of assessment that is not valid in other realms, that might simply be dismissed as sentimentalism or utopianism or parochialism. And, of course, these charges can be levelled at literature too. However, after recognition of the supplement literature provides to social discourse (Angenot 219), our view of that social discourse may change; we may read the social discourse differently. When a country's federal government engages in schoolyard name-calling regarding a review process that approves over 99 per cent of applications, we may read that as contradicting a purported focus on scientific facts, as an attempt at persuasion unconnected from a concern with the truth. It might suggest that what we have heard up to now has been mostly "bullshit," and it is time to talk about matters of concern.

Irrational **Oil**

Ethos, Extraction!, *Elder Brother, and Free Speech*

IN THIS CHAPTER, I will, as an alternative, or counterpart, to the complexity of matters of concern, try to think through an explicitly irrational response to debates around bitumen extraction through contrasting readings of Elder Brother in Warren Cariou's "An Athabasca Story" and the speaker in "From the Bottom of the Pit" in *Extraction! Comix Reportage*. First, though, I want to continue the investigation into the role of *ethos* in the debate over bitumen begun in the previous chapter by considering the truth-telling figure of the *parrhesiastes*.

The Truth: *Parrhesia*, Character, and Risk

Ethos and irrationality come together in the parrhesiastes because it is, at least in part, through the "irrational" act of putting him/herself at risk that the parrhesiastes establishes credibility. While many speakers in these debates attempt to establish persuasive ethos, few attain the status of parrhesiastes because their speech does not put them at risk. The counterfactual realm of fiction offers one space where such risk can be imagined and speakers can attain the status of parrhesiastes, thereby confronting readers with a moment of self-awareness in regard to an ethic of self-care. As Michel Foucault explains, "the decisive criterion which identifies the parrhesiastes is not to be found in his birth, nor in his citizenship, nor in his intellectual competence, but in the harmony which exists between his *logos* and his *bios*" (106). This harmony between words and actions, between beliefs and life, extends, as in the life of Socrates, to the point of dying for one's beliefs.[1] Further, "the target of this new *parrhesia* is not to persuade the Assembly, but to convince someone that he must take care of himself and of others; and this means that he must *change his life*" (106, emphasis original). The realm of fiction provides one arena for such persuasion.

Increasingly, industry and government are working to defend the moral character of bitumen extraction (by emphasizing economic and social benefits and environmental responsibility); critics work to expose this narrative's flaws. As we have seen in previous chapters, facts have been marshalled by both sides to appeal to society's collective *logos*, and a range of striking images and sentiments have appealed to our *pathos*, but the debate, in my view, is more and more about ethos. As Aristotle notes, "character contains almost the strongest proof of all" (75) because "we more readily and sooner believe reasonable men [*sic*] on all matters in general and absolutely on questions where precision is impossible and two views can be maintained" (74). In such cases, the inadequacy or incompleteness of logical appeals give way to the ethical—the perceived character of the speaker. We are persuaded not by those who present the best facts (most of us are not in the position to judge which facts are better) but by those who appear to us as most trustworthy. This helps explain why criticisms about development are now met with public relations.[2] All sides in the debate are working to establish their character, to present themselves as on the side of truth.

However, one way, perhaps, of distinguishing among these claims is through the concept of parrhesia. In his 1983 lectures on "Discourse and Truth" at Berkeley, Foucault explored the Greek concept of parrhesia, saying,

> Etymologically, *parrhesiazesthai* means "to say everything"...The one who uses *parrhesia*, the *parrhesiastes*, is someone who says everything he has in mind: he does not hide anything, but opens his heart and mind completely to other people through his discourse. In *parrhesia*, the speaker is supposed to give a complete and exact account of what he has in mind so that the audience is able to comprehend exactly what the speaker thinks. The word *parrhesia*, then, refers to a type of relationship between the speaker and what he says. For in *parrhesia*, the speaker makes it manifestly clear and obvious that what he says is his own opinion. And he does this by avoiding any kind of rhetorical form which would veil what he thinks. Whereas rhetoric provides the speaker with technical devices to help him prevail upon the minds of his audience (regardless of the rhetorician's own opinion concerning what he says), in *parrhesia*, the *parrhesiastes* acts on other people's minds by showing them as directly as possible what he actually believes. (12, emphasis original)

He later notes that "in the Greek conception of *parrhesia*...there does not seem to be a problem about the acquisition of truth since such truth-having is guaranteed by the possession of certain *moral* qualities: when someone has certain moral qualities, then that is the proof that he has access to truth—and vice versa" (15). The "proof," if there is one, Foucault writes, "is his *courage*. The fact that the speaker says something dangerous—different from what the majority believes—is a strong indication that he is a *parrhesiastes*" (15). Thus, the parrhesiastes establishes his/her ethos not through rhetorical appeals or by evidence of specialized knowledge, but by putting at risk, through speech, his or her self.[3] This danger is an essential element in *parrhesia* and, therefore, "the king or tyrant generally cannot use *parrhesia*; for *he* risks nothing" (16).[4] The danger the parrhesiastes is in, further, is of a particular type: it "always comes from the fact that the said truth is capable of hurting or angering the *interlocutor*...The function of *parrhesia* is not to demonstrate the truth to someone else, but has the function of *criticism*: criticism of the interlocutor or of the speaker himself" (17). Lastly, "in *parrhesia*, telling the truth is regarded as a *duty*. The orator who speaks the truth...is *free* to keep silent. No one forces him to speak, but he feels that it is his duty to do so" (19).

Foucault goes on to trace the evolution of the word in relation to rhetoric, politics, and philosophy. He notes that "in the Socratic-Platonic tradition, *parrhesia* and rhetoric stand in strong opposition...The continuous long speech is a rhetorical or sophistical device, whereas the dialogue through questions and answers is typical for *parrhesia*" (20). Thus, while the texts under discussion here lay some claim to parrhesia, most retain rhetorical attempts at persuasion, attempts to demonstrate "the truth" to someone else. The long speech in "From the Bottom of the Pit" would, in Foucault's terms, be a sophistical device while Elder Brother's dialogue with the operator in "An Athabasca Story" is more parrhesiastic, although Elder Brother is presented as naïve rather than duty-bound and courageous.

However, these divisions are not clear-cut, as Foucault sees the parrhesia being absorbed into rhetoric at the beginning of the Roman Empire. In *Institutio Oratoria*, Quintilian discusses rhetorical figures that are "adapted for intensifying the emotions of the audience...[which] he calls by the name *exclamatio* (exclamation)" (21); in this category Foucault describes

> a kind of natural exclamation which...is not "simulated or artfully designed." This type of natural exclamation he [Quintilian] calls "free speech" which, he tells us, was called..."*parrhesia*" by the Greeks. *Parrhesia* is thus a sort of "figure" among rhetorical figures, but with this characteristic: that it is without any figure since it is completely natural. *Parrhesia* is the zero degree of those rhetorical figures which intensify the emotions of the audience. (21)

As discussed in the previous chapter regarding the Northern Gateway debate (which also applies, of course, to TransCanada's Keystone XL and Energy East proposals, Enbridge's Line 9 reversal, and the Kinder Morgan Trans Mountain Pipeline expansion), it is impossible to guarantee that the pipeline will leak and it is impossible to guarantee that it won't. Therefore, Enbridge, the Canadian Association of Petroleum Producers (CAPP), the Joint Review Panel, and the provincial and federal governments work to establish their moral character as defenders of the environment who will "balance" ecology in a reasonable and practical manner with economic growth. The opposition attempts to make alternate associations. Both sides lay claim to "the facts" as a way of showing that they are operating outside of "rhetoric." However, the facts, as argued in the previous chapter, are inadequate and do not offer unmediated access to "truth." Most of the speakers in this debate do not attain the status of parrhesiastes because they do not risk anything through their speech. Elder Brother, though, as we will see, through his unfiltered speech—he simply says or exclaims what he thinks—does find himself in danger and, in so speaking, reveals the contours of the hegemony in place.

The Whole Truth: Counterfactuals As Necessary Supplement to "Facts"

Although it has been argued that establishing an extreme position on the fringes of a discourse can help to shift the centre in that direction, this maneuver is available to all sides in a debate.[5] As mentioned in Chapter 3, "In a world where everything is neutralized...the status quo wins" (Corlett 197). In such a world, where "bullshit" predominates, is it reasonable to expect a parrhesiastes to appear or to expect that we would recognize one if he or she did? Literature offers one space in which the figure of the parrhesiastes can appear as a model for readers. It is important, here, though, to consider Foucault's distinction between political parrhesia and parrhesia in the care of the self. For the first, Foucault considers Isocrates's view that, in Athens of the 4th century BC, democracy led to a situation in which the people would only listen to those who said what they wanted to hear:

> The difference between the good and the bad orator does not lie primarily in the fact that one gives good while the other gives bad advice. The difference lies in this: the depraved orators, who are acceptable by the people, only say what the people desire to hear. Hence, Isocrates calls such speakers "flatterers"...The honest orator, in contrast, has the ability, and is courageous enough, to oppose the demos. He has a critical and pedagogical role to play which requires that he attempt to transform the will of the citizens so that they will serve the best interests of the city. This opposition between the people's will and the city's best interests is fundamental to Isocrates's criticism of the democratic institutions of Athens. And he concludes that because it is not even possible to be heard in Athens if one

> does not parrot the demos' will, there is democracy—which is a good thing—but the only parrhesiastic or outspoken speakers left who have an audience are "reckless orators" and "comic poets"...Hence real *parrhesia*, *parrhesia* in its positive, critical sense, does not exist where democracy exists. (82)

By contrast, Socrates "is able to use rational, ethically valuable, fine, and beautiful discourse; but unlike the sophist, he can use *parrhesia* and speak freely because what he says accords exactly with what he thinks, and what he thinks accords exactly with what he does. And so Socrates—who is truly free and courageous—can therefore function as a parrhesiastic figure" (101). However, "Socratic *parrhesia* differs from political *parrhesia* in a number of ways. It appears in a personal relationship between two human beings, and not in the *parrhesiastes'* relation to the *demos* or the king. And in addition to the relationship we noticed between *logos*, truth, and courage in political *parrhesia*, with Socrates a new element now emerges, viz., *bios*" (101). Ultimately, "The aim of this Socratic parrhesiastic activity...is to lead the interlocutor to the choice of that kind of life (*bios*) that will be in Dorian-harmonic accord with *logos*, virtue, courage, and truth" (101). Socratic parrhesia, then, operates in the realm of what Foucault calls "self-care" rather than politics and can usefully be connected to the reading of literature, as we will see below.

Foucault, of course, elsewhere in his writing, takes a rather different view of truth and is often aligned with the postmodernist view that there is no way of distinguishing "truth" from "bullshit." However, in taking up this tension in Foucault's thought, Zachary Simpson writes,

> "Fictions" are, as [Foucault] says, "experiments" in truth. And, as such, the line between fiction and truth is easily blurred: fiction produces the same effects as true discourse, and, epistemically, stands on the same plane as that which is held to be true. Fiction effectively holds the same epistemic weight as truth. Both are produced and productive; both can actively frame discourse with respect to bodies and societies. Yet fiction holds a decisive advantage over what we regard as truth in that it

> constructively imagines an alternative interpretation of the present that is not yet available.

This idea of "alternative interpretation" is another way of talking about counterfactuals, and the ways in which truth and fiction are both "produced and productive" relates to the ways in which facts and counter-facts interact. Simpson continues,

> In this way, fiction has a proleptic function, calling forth and enacting a new reality through its pronouncement. Like *parrhesia*, fiction is an enactment of a truth within a present reality. Fictioning is an active process of bringing about the same effects as truth, though they may not currently exist. Seen this way, "truth" is that which has effects in the present, while "fiction" is that which accurately reflects the present while having effects in the future. Neither are "true" in any sense other than their functioning within power relations. (7)

This functioning of "truth" and "fiction" within power relations suggests the ways in which free speech can intervene within ideology to create change: fiction can enact change by expressing the present state of an individual's heart or mind to a reader such that the reader's life is changed, and, thereby, the future is changed.

Sourayan Mookerjea notes the importance of utopian thinking to oppose existing ideology: "not only does ideology involve identification, mystification, and depoliticization, but ideology's utopianism—insofar as all ideology allegorically and intertextually projects some discourse of community—leaves any ideology open to the possibility of repoliticization" (247). Such repoliticization can occur through the use of downward counterfactuals that speak the truth about actually existing ideology's failings and the weaknesses in its upward counterfactual utopianism. Simpson argues that

> fictive truths serve a critical function in Foucault's later thought and can be formidably linked to both his aesthetics of existence and his work with

> Hellenistic practices of parrhesia. Fictions allow for the creation of new truths which can form points of resistance within power relations. They also serve to produce an interference with the present and to reconstitute future relations in the midst of that dissonance. (8)

As Mookerjea notes,

> the referent of community discourse is nowhere existent, is precisely nothing and nowhere, and this is what makes discourses of community Utopian. It is this aspect of community discourse's ideological Utopianism, its negative referentiality or its poetics of idealization, that is the condition of possibility for the negation of any ideological symbolic universe that exists historically. The Utopian poetics of community discourse allow us to negate and criticize what is the case: the present system of domination and exploitation. (248)

This poetics of community that does not exist, that is *on the way*, gives fictional literature a power that the real, enacted, historical events of ideology cannot protect against (though ideology attempts to do so through its own utopian poetics of community—upward counterfactuals).

In "An Athabasca Story," the utopian ideology of the community created through bitumen extraction is negated through Cariou's fictionalization of that community's treatment of Elder Brother. Robert Alexander Innes, in *Elder Brother and the Law of the People* states that "the Elder Brother stories explained the rules and expectations for normative behaviour" (38) but also quotes Donald Auger's claim that "stories act to connect our 'normative system to our social constructions of reality and to our vision of what the world might be'" (38). This relationship between normative behaviour and the potential for what the world might be suggests a complicated relationship between facts and counterfacts in Elder Brother stories.

Clinton N. Westman describes the Cree trickster as follows:

> Due to lack of self-knowledge and inability to control his appetites, the Trickster (who for the Cree resembles his namesake, the voracious Grey Jaybird) deceives humans and animals; but frequently cannot avoid the

> consequences of his actions. The Trickster's adventures are earthy indeed, sometimes involving his disobedient anus and enormous penis, which can take on a will of their own. It is easy to imagine the Trickster's sexual and culinary adventures (as well as their sometimes unpleasant consequences) in relation to the oil sands projects...as gigantic devices penetrate the earth, and water and land are devoured whilst foul waste belches out in their place. (225)

While Elder Brother's, or wîsahkêcâhk's, lack of self-knowledge contrasts the parrhesiastes' congruence between words and actions, it also functions like the Socratic *doctrina ignorantia* for the reader/listener: if true wisdom is knowing that you don't know and this state can prod you on to question your assumptions, then Elder Brother's ignorance also exposes the ignorance of those who presume to know (and, by extension, the purported knowledge of the reader/listener). In the example of Cariou's "An Athabasca Story," we find Elder Brother hungry and alone in an unfamiliar land, looking for assistance from his relations. He finds, instead, "a vast expanse of empty land" (70), an alien landscape devoid of life; he has walked into a bitumen mine. Along with this lifelessness, as we will see, he discovers as its counterpart a world in which the rules governing normative behaviour are very different from what he is used to, a world in which acquisitiveness has replaced sharing and selfishness has replaced community. Elder Brother's actions, and his ignorance, in "An Athabasca Story," serve to remind readers of the normative rules of Cree culture and to point to an alternative idea of what the world might be.

...And Nothing But the Truth?

While he might be dismissed as taking the role of the "comic poet," Elder Brother, rather, in what we might term a satirical Socratic dialogue, raises the audience's self-consciousness and points both to the need for harmony between words and actions and to the consequences of living in a world without such harmony. Innes argues that "as the most important character in Cree/Ojibwe stories, Elder Brother is more than just a trickster; he is really the Cree/Ojibwe cultural hero" (34). He goes on to note, "The failure to follow prescribed regulations could, according to what happens to Elder Brother in the stories, result in severe negative consequences. Conversely, adhering to positive behaviour that Elder Brother displays was seen as the ideal that all should strive to attain" (37). The *Extraction!* speaker attempts to operate through political parrhesia as a "reckless orator" but without obviously risking anything. We never know if his words and actions accord. In "An Athabasca Story," we see Elder Brother speak of the importance of kinship, but, in the end, he acts selfishly. Thus, he embodies "the internal clashes attending confrontation between instinct and reason, emotion and thought" characteristic of trickster narratives (Innes 29).[6]

The graphic book, *Extraction! Comix Reportage* attempts to present the environmental consequences of global mining practices in a new way to counter the status quo. It presents journalistic research combined with graphic art to raise awareness about the depredations of the global mining industry. In discussing the availability of images that construct a counter-frame to the dominant industry and government narrative about bitumen, Mike Gismondi and Debra Davidson claim, "If an image says more than a thousand words, and we think they do, then the tar sands tale has been told to hundreds of millions of people, who, drawn to them like a lighthouse, may just see a common future and basis for global citizenship" (171). This quotation could serve to describe what the editors of *Extraction! Comix Reportage* hope to achieve with their book, which looks into Canadian mining companies operating in Guatemala, India, Quebec, and Alberta. The editors see the book as performing a parrhesiastic function, and, indeed, many of the people represented in the book, and the journalists telling their stories, put themselves at risk to speak out against the actions of global mining companies.

The final chapter—"From the Bottom of the Pit"—addresses bitumen. In it, a speaker climbs onto a soapbox to expose the industry, and, as he is speaking, a large crowd gathers to listen. This, it seems to me, is precisely what is not happening with the debate about bitumen or the extraction issues that are highlighted in the book more generally. Although people are speaking out against mining practices, we inhabit democratic institutions in crisis, as Foucault discusses in relation to Isocrates's "Ariopagiticus" where the positive characteristics of democracy ("liberty...happiness...and

FIGURE 10: Extraction! Comix Reportage, *page 91.*

equality in front of the law" [83]) are debased: "Democracy has become lack of self-restraint[,]...liberty has become lawlessness[,] happiness has become the freedom to do whatever one pleases[,]...and equality in front of the law has become *parrhesia*. *Parrhesia* in this text has only a negative, pejorative sense" (83). This debasement exists in a system where everyone can say everything they want and "it is not even possible to be heard...if one does not parrot the *demos'* will" (Foucault 82). In such a situation, freedom of speech, letting everyone say everything they want, creates the neutralization Corlett describes.

The use of pictures in *Extraction!* does not, on its own, enable the book to install a successful counterframe, as Gismondi and Davidson suggest may be possible. Industry and government also combine pictures and facts to make their points and are able to reach a much larger audience. The facts we hear wash over us as part of the media wave we inhabit. When someone gets up on a soapbox, someone else gets up on another and refutes the "facts" of the first. Or, if the speaker is able to successfully take the role of political parrhesiastes, the dominant discourse has become adept at marginalizing those counterdiscourses.[7]

While most of the comix is focused on introducing readers to bitumen and showing the ecological consequences of its extraction, the speaker also shows how environmental groups are bought off by big oil via the Pew Trust, a US charitable organization created by the Pew family, founders of Sun Oil and the Great Canadian Oil Sands, forerunner of Suncor. "In November 2005," the speaker declares, "[the] Pembina [Institute] produced a report called 'Counting Canada's Natural Capital: Assessing the Value of Canada's Boreal Ecosystems'" (102):

> The sponsor of this "natural capital" report is the Canadian Boreal Initiative (CBI), self-described as "an independent organization working with conservationists, First Nations, industry and others to link science, policy and conservation activities in Canada's boreal region." This "organization" does not have a Board of Directors, is not registered as a non-profit and does not have a charitable number...As noted on CBI's own website, its sole funder is actually the Philadelphia-based Pew Charitable Trusts. (103)

He goes on to describe how Pew funnels money into Canada despite not being a legal entity in Canada: "Pew first transfers money to the North American headquarters of Ducks Unlimited in Nashville, Tennessee, which is transferred to its Canadian headquarters in Winnipeg. Ducks Unlimited...then recycles 'boreal' money not only back to CBI, but also to other Canadian environmental organizations" (104). He concludes by asserting that a "pattern of environmental organizations adopting a docile 'low-hanging fruit' strategy soon after being bankrolled by Pew has been thoroughly documented" (105). In short, Pembina via CBI and Pew has been bought off with oil money. Taking a more moderate position, Gismondi and Davidson say that although regional environmental groups like Pembina "have found themselves in the precarious position of depending on funding from the very corporations they criticize," they also "have provided some of the most valuable counter-information into local debates, but by and large have adopted 'consensus building' tactics, designed to ensure a seat at the table, rather than turning the tables" (177).

Ezra Levant, by contrast, in *Ethical Oil*, attacks the Pembina Institute for its anti-bitumen lobbying efforts and argues that the money it raises via the Pembina Foundation for Environmental Research and Education, a registered charity, is raised under false pretenses since it is then transferred to the Pembina Institute, an advocacy group. For Levant, "Pembina has to raise an enormous amount of money every week just to meet payroll" (159), and that, like other environmental charities, is their base motivation for everything they do. In 2012, Geoff Dembecki reported that the federal government was implementing limits on charities' ability to use funds in these ways. By 2014, Dembecki's concerns seem to have been borne out with several reports suggesting that the focus of the Canada Revenue Agency's charity audits have been targeted at left-leaning charities whose goals conflict with those of the Conservative government (Nuttall; Cheadle; Mayer).

On the one hand, Pembina is co-opted by the right; on the other hand, they are co-opted by the left, and both left and right criticize their bias. Given this information, what do we make of Pembina's ethos? Both extremes on the continuum of this debate criticize Pembina's ethos, their ability to serve as a parrhesiastes. This is just one example of how someone seeking to cultivate an informed opinion about bitumen is faced with a dizzying array of claims and counterclaims. The consequence, I would suggest, is either paralysis (it is impossible to find anyone who is trustworthy, whose words and actions are congruent) or a deliberate choice to accept one view and dismiss the other (the words of so-and-so support my actions and desires, so I will accept that position). This latter may be the more likely of the two. As Cariou suggests, "Who can deny that some forms of thought create a noxious atmosphere, a stink, sometimes subtle and other times overwhelming? We all believe this about the people we disagree with, the ideologies we hate" ("Tarhands" 28). However, if ideological opponents are unable to hold their noses and engage in the mediating ground of rhetoric, then what hope is there for democracy, which depends on a public that can be addressed? Conny Davidsen speaks of the "reason-giving" requirement of deliberative democracy, but, if citizens have already dismissed their opponents' ethos, they will not listen to their reasons: we turn away from them as from a bad smell.

Keith Grant-Davie, in "Rhetorical Situations and Their Constituents," argues that the "notion of the universal audience [which Chaim Perelman defines as] an audience 'encompassing all reasonable and competent men [*sic*]'...creates a forum in which debate can be conducted" (271). Without such a forum, is the potential for change limited to the individual *bios*, to parrhesiastic self-reflection? James Porter writes,

> A discourse community shares assumptions about what objects are appropriate for examination and discussion, what operating functions are performed on those objects, what constitutes "evidence" and "validity," and what formal conventions are followed....Some discourse communities are firmly established, such as the scientific commnity, the medical profession, and the justice system...In these discourse communities..."a speaker must be 'qualified' to talk; he has to belong to a community of scholarship; and he is required to possess a prescribed body of knowledge...[This system] operates to constrain discourse; it establishes limits and regularities...[it establishes] who may speak, what may be spoken, and how it is to be said; in addition [rules] prescribe what is true and false, what is reasonable and what foolish." (39)

FIGURE 11: *"What we stand to lose" poster.* [BC Teachers' Federation]

When the discourse community is "the universal audience," though, the ethos is full of competing functions and indefinite boundaries, and the competition to define the boundaries of who may speak, what may be spoken, and how it is to be said is unending, and can be part of the hegemonic strategy to maintain the status quo: attacking your opponent's character can serve to perpetuate doubt and prevent coalition building.[8] We see these attacks employed in Joe Oliver's letter and his comments on the BC Teachers' Federation's development of materials on the Northern Gateway Pipeline, which stress the risks associated with a major oil spill (O'Neil, "Teachers" A7). At the same time, as Daniel Coleman argues, reading, especially reading fictional literature, can, paradoxically, through its solitariness, take the reader beyond the self and beyond common sense, hegemonic understandings of the world (13). The parrhesiastes manages to say what otherwise could not be said, to speak beyond the limits of the discourse community.

Grant-Davie makes the commonplace connection between rhetoric and war, and Oliver's language certainly upholds this comparison:

> On the battlefield, one side's ability to select the ground to be contested has often been critical to the outcome of the engagement. In the same way, rhetors who can define the fundamental issues represented by a superficial subject matter—and persuade audiences to engage in those issues—[are] in a position to maintain decisive control over the field of debate. (267)

In this sense, the separation of bitumen from the complex ecology in which it is located is the "superficial subject matter"; it is the symptom, not the disease, the battle, not the war. As I have been arguing throughout *Unsustainable Oil*, the extraction and refining of hydrocarbons is a product of the culture we have created; if we want to change our relationship to hydrocarbons, we need to change our culture. Industry and government rhetoric has the advantage of the status quo's inertia, the dangers of potential change, and the connection to the upward counterfactual of utopian "sustainable development" as a narrative of progress. This rhetoric continues to define the ground to be contested despite attempts to shift to different battlefields, to question the assumptions of the discourse community and its ideology.

Frédéric Dubois, one of *Extraction!*'s editors, notes that one of the limits of the book is that it does not "elaborate on the escalation of 'counter-insurgency' tactics used by mining corporations to delegitimize their opponents" (16). In the bitumen debate, this delegitimizing is seen in many documents examined in this book—regular op-eds by the Canadian Association of Petroleum Producers, media campaigns like ethicaloil.org, and the co-optation of environmental groups pointed to in "From the Bottom of the Pit." Dubois asserts, "There are many ways to affect the course of events. One is by extracting stories and sharing them. This is what we have done" (17). Marc Tessier, another of the editors, quotes George Orwell, "In a time of universal deceit, telling the truth is a revolutionary act" (118), but, as Foucault implies, the telling and the sharing are only revolutionary if there is an audience who will listen, who can recognize what they hear as the truth, and who will then tell and share in turn: this listening and recognition depends on the speaker's ethos. When the soapbox speaker concludes, "if we all voice our concerns. We will be heard!" and the audience applauds him, it feels—especially following the clear disregard for those who question corporate actions evidenced by Goldcorp and Alcan discussed in *Extraction!*'s previous chapters—more like wishful thinking than the beginnings of a revolution.

The speaker, a fictional parrhesiastes, inserted into a "factual" journalistic account of bitumen mining, does not engage in dialogue, but presents a long speech. Ultimately, it is up to the reader to decide if he is engaged in parrhesia or sophistry. Regardless of how we decide, Oliver and *Extraction!* both seek to define the issues to be debated. The tactics employed by Oliver are meant to restrict democracy by defining who may speak; this is a version of what John Dryzek calls "Leave It to the Experts: Administrative Rationalism" (63). *Extraction!* suggests that we should leave it to the people, what Dryzek call "democratic pragmatism" (117). However, it seems to me that these two positions do not speak to the same discourse community, let alone a universal audience. Instead of engaging in the type of collaborative debate required for an "ethical argument" as Theresa Enos defines it, where "writer and reader are able to meet in the audience identity the writer has created within the discourse" (qtd. in Grant-Davie 268), battle positions continue to be entrenched. Rhetors speak to the people who already agree with them and occasionally lob a grenade at the other side, while those in the middle tune out or say everything they have to say in other forums (like posting comments on YouTube or news forums). As Grant-Davie notes of the "ethical argument," "In these kinds of argument, establishing acceptable issues would seem to be an essential stage, creating an agenda that readers can agree to discuss" (268). What are the acceptable issues in bitumen development? The polarization of the debate leads to less and less mediatory ground on which opponents can meet to discuss anything.[9]

In its best sense, rhetoric suggests a coming together, a meeting of minds in a mutually beneficial undertaking: as Enos describes, "the aim is not victory over the opponent but a state of identification...created within the discourse" (qtd. in Grant-Davie 268). In its worst sense, rhetoric suggests manipulation, a win-at-all-costs use of discourse to deceive an audience into thinking in a particular way. Lloyd Bitzer defined "rhetorical audience" as consisting of "only...those persons who are capable of being influenced by discourse and of being mediators of change" (qtd. in Gorrell 395). This raises the question of whether, when it comes to bitumen extraction and the various pipeline debates, there is an exigence, a rhetorical opportunity for change within the rhetorical audience—those capable of bringing about change—especially when the government seems intent on limiting opportunities for citizens to access regulators (as in the use of its omnibus budget Bill C-38 to, among other things, "streamline" the process for environmental reviews and weaken the habitat protections of the Fisheries Act). It seems to me that very often in this debate the various participants are already more or less dogmatic adherents to their positions. Indeed, nearly every statement made from a particular perspective generates a flurry of opposition activity to debunk it and perpetuate relativist doubt. This doubt, I think, serves the status quo as it gives people another excuse not to think about their implications in the oil economy or to rationalize those implications as better than others. As Stoekl says, "we know enough to want not to know" ("Unconventional Oil" 40).

"You Can't Handle the Truth": Relationship and Irrationality

In "Tarhands: A Messy Manifesto," Cariou "sets aside any attempt to make reasoned arguments about conservation or regulation, and instead embraces irrationality as the last possible mode of engagement with a contemporary public that will no longer listen to reason" (17). What follows here attempts to think through the following questions in relation to Cariou's "An Athabasca Story": Should we embrace irrationality as a response to this impasse? Should we cultivate an ethos of irrationality? And, if so, how does such a response differ from the irrationality of bullshit considered in the previous chapter? Or, given the common sense response to arguments for, or from, irrationality—that they will lead to the marginalization of that position—does Cariou's use of the figure of Elder Brother offer an alternative form of (ir)rationality? Westman suggests that "we, as human beings, are putting ourselves in danger by forgetting what we know about community and relatedness, and by accepting this deeply distorted, sociopathic, corporate logic of growth and consumption as common sense" (224). As James H. Kavanagh shows, it is precisely through "common sense" that ideology operates: "It is much more effective—and cheaper—to put 'You can't fight City Hall,' or 'the poor will always be with us,' or 'every revolution just leads to worse tyranny' on everyone's lips than to put all the cops on all the corners that would be necessary to confront any determined struggle of the poor and homeless against the social system that produces poverty and homelessness" (309). We might, then, understand Cariou's call for an "irrational response" to bitumen extraction as an attempt to expose the flaws in the "rational" and "common sense" logic of capitalism, a move to "uncommon sense." This "logic" of capital is countered through the "irrationality" of relatedness in Cariou's "An Athabasca Story."

Simpson elaborates on Foucault's point about the key characteristic of the parrhesiastes being the harmony between *logos* and *bios*: "On one level, this is the ability, as Foucault recommends, to 'show that there is a relation between the rational discourse, the logos, you are able to use, and the way that you live,' a harmony between discourse and action" (10). "An Athabasca Story," in these terms, does not seem to me to enact Cariou's call in "Tarhands" for an irrational response to bitumen extraction, but it does work to reveal the irrationality of our cultural relationship to bitumen; it reveals the lack of harmony between our discourse and our actions through the trickster figure of Elder Brother.

Cariou puts the Cree figure of wîsahkêcâhk, or Elder Brother, in the bituminous region of Northern Alberta:

> One winter day Elder Brother was walking in the forest, walking cold and hungry and alone as usual, looking for a place to warm himself. His stomach was like the shrunken dried crop of a partridge...
>
> Where will I find a place to warm myself? he wondered. Surely some relations will welcome me into their home, let me sit by the fire.
>
> But he walked for a very long time and saw none of his relations. (70)

Eventually, after walking so far west that he no longer recognizes the land, and just as "he was nearly at the point of...giving up, Elder Brother thought he smelled something. It was smoke, almost certainly, though a kind of smoke he'd never encountered before" (70). This smoke is coming, it turns out, from a bitumen mine. Despite the stench, "worse than his most sulfurous farts, the ones he got when he ate moose guts and antlers" (70), Elder Brother, in his desperation, enters the mine, where he encounters the driver of a heavy hauler: "Oh my brother, my dear relation, Elder Brother said. I'm very cold and hungry and I was hoping...to come and visit you in your house" (71). After the operator's initial confusion about how Elder Brother came to be there—"Who in hell are you? Where's your machine?" (71)—the operator asserts, "Well you'll be a lot worse than cold...if you don't get the hell out of my way and off this goddamn property" (71). After threatening Elder Brother with calling security, having him imprisoned, or running him over, Elder Brother thinks "*that* was rude...but he tried not to betray his disappointment. This man talked as if he had no relations at all" (73). Before he steps aside, though, he asks the man what they are doing with the earth: "Burning it" (72), the man replies.

In "The Ethical Space of Engagement," Cree legal scholar Willie Ermine argues, "The treaties still stand as agreements to co-exist and they set forth certain conditions of engagement between Indigenous and European nations" (200). Although the Treaty 8 commissioners noted, in their report to the

FIGURE 12: Extraction! Comix Reportage, *page 97.*

minister of Indian affairs, that they were "so hurried [in their journey] that [they] were not in a position to give any description of the country ceded which would be of value," they pointed out that "the country along the Athabasca River is well wooded and there are miles of tar-saturated banks" (Treaty 8). At the same time, the commissioners went to some pains (as did Charles Mair, discussed in Chapter 2) to assure the First Nations that the "same means of earning a living would continue after the treaty as existed before it" (Treaty 8). The treaty itself reads: "Her majesty the Queen HEREBY AGREES with the said Indians that they shall have right to pursue their vocations of hunting, trapping and fishing throughout the tract surrendered...subject to such regulations as may from time to time be made by the Government of the country...and saving and excepting such tracts as may be required or taken up from time to time for settlement, mining, lumbering, trading or other purposes" (Treaty 8). The First Nations of Treaty 8 maintain their oral history around the signing of the treaty, which views the agreement as continuing "as long as the grass grows, the sun shines, and the rivers flow." Along with the signatories of Treaties 6 and 7, they have developed a position paper calling for changes to the Alberta government's process of consultation, because it has not adhered to the spirit of the treaty. It notes that "the Treaty is not a finished land use blueprint" and that "consultation is an ongoing process" (Treaty 8 Alberta Chiefs). Cariou's story reveals a contradiction between a continuation of pre-treaty means of living and taking up tracts of land for the purpose of burning them.

I suspect Cariou, in emphasizing the burning of the earth in his story, is thinking of an encounter between Crowfoot and government officials following the negotiations of Treaty 7 in Southern Alberta: "Crowfoot took some earth in one hand and, throwing it into the fire, said the earth would not burn. He then said that if he took money and threw it in the fire, it would burn and disappear. Finally, he said that he would rather keep the earth because it will not burn" (Treaty 7 Elders 117). Treaty 7 elders remember this story as evidence that the First Nations' understanding of the treaty was not one of ceding the land. In "An Athabasca Story," Elder Brother responds to the assertion that people are burning the earth with,

"Burning. But earth doesn't—" (72) before he is cut off by the man: "This stuff does...You really are a moron, aren't you? It's very special dirt, this stuff" (72). If I am right in connecting these narratives, Cariou's story raises several questions: If the earth will burn, how would Crowfoot have responded? If burning the earth will create money, would his response have been different? What should our response be? If we see Elder Brother's role as a kind of parrhesiastes, albeit one whose *lack* of reason reveals what we should do, how should we change our lives?

In "Dances with Rigoureau," Cariou considers the history and associations of the Métis trickster *rigoureau* in relation to his own family experiences and the broader culture. He notes its origins as a *Michif* word, derived from the French *loup-garou*, a werewolf figure in French folktales (159). He writes,

> Any recognition that humans are part animal would short-circuit the colonial paradigm of maximizing profit through the utilization of land and animals. I think there are also some other related reasons that the loup-garou is virtually always a symbol of terror in European cultures. The loup-garou, like the werewolf, is obviously a manifestation of a fear of giving into irrationality. But it also represents a fear of an unstable humanity, a humanity that is subject to periodic fluctuations—and also a humanity that is hybrid. (160)

While Elder Brother is a trickster figure with a different genealogy than the *rigoureau*, he also points to an unstable and hybrid humanity, a humanity that is both rational and irrational, animal and non-animal, and, as such, points to truths that we may prefer to ignore. In "An Athabasca Story," it is anger, rather than fear, that Elder Brother's hybridity evokes. As the man's anger in response to Elder Brother's questions makes clear, maximizing profit through utilization of land and animals is not rational. The man cannot rationally explain or justify what is happening and quickly resorts to threats and name-calling.[10] Elder Brother exposes the constructed nature of the ideology at work—the man's "common sense" answers do not persuade Elder Brother, who stands outside the "rationality" of ideological state apparatuses—and the man quickly invokes the repressive apparatuses—security, imprisonment—that stand behind ideological explanations (Althusser).

However, rather than the duty-bound action we might expect from a parrhesiastes, Elder Brother reveals his appetitive nature, that his *logos* about relatedness does not correspond to his *bios*: his words and actions are not congruous. His misery leads him to grab some of the magical dirt for himself and to justify the action: "Since the man and his company were so rude, they deserved to have their precious dirt stolen" (5), he thinks. When he thrusts his arms into the earth a voice cries out:

> Ayah!...What are you doing, Elder Brother?
>
> Sssshhhhhhhh, he answered. I'm taking what's mine.
>
> And he reached deeper and deeper into the ground, spreading his fingers as wide as he could in order to hold the largest armload of dirt...
>
> Elder Brother, you're hurting me! the voice cried out.
>
> Not nearly so much as they are, he said, and with that he began reaching in with his other arm, tunneling in with his fingers, opening his arms wide in a desperate embrace. (73)

Despite Elder Brother's attempt to rationalize his own actions as relatively minor in relation to the scale of bitumen mining and the reality of his need, this selfish act is punished:

> But when Elder Brother tried to lean back and lift the great clump of dirt out of its place, he discovered that he had no leverage...He grunted and panted, flexed again, shimmied his buttocks for extra oomph. However it didn't make a bit of difference...
>
> Instinctively he called out to the voice that had spoken...Help me! I'm sorry I didn't listen to you. I'll leave now without taking anything at all. (73)

The voice, though, does not answer, and Elder Brother is stuck in the ground for two days before he is dug out by the machines, dumped into the back of a truck, and hauled to the processing plant. In the end we are told,

> Of course Elder Brother can't die, luckily for him. Or perhaps not so luckily. He's still alive even now, after everything he's been through. It's true that people don't see him much anymore, but sometimes when you're driving your car and you press hard on the accelerator, you might hear a knocking, rattling sound down deep in the bowels of the machine. That's Elder Brother, trying to get your attention, begging you to let him out. (75)

This conclusion suggests that humans are lucky in our ability to escape life through death, but also, more significantly, that our actions cause pain to our relations who cannot die, to those who will be left to continue on after the anthropocene ends. The voice of the earth cries out when Elder Brother attempts to dig up bitumen even though the amount he takes is relatively small. The sound of the engine knocking in "your" car is the sound of Elder Brother begging to be let out. Stoekl writes of the automobile as "the empty signifier at the summit of fossil fuel modernity" (*Bataille's Peak* 184). He continues,

> All is mediated through the automobile: everyone derives the meaning of their lives through it: as a status marker, as a simulacrum of freedom of movement and consumption (David Brooks's utopia), as the timelessness of a religion shared by all. Virtually everyone in American society works as hard as they do to pay for their cars: it is their major investment, the acquisition that justifies and represents human labor ("drive to work, work to drive"). And, as common denominator, it is the transcendence of labor, of meaning: sitting in traffic jams, the driver does nothing, just sits, and in this way lavishly neutralizes the labor devoted to purchasing the vehicle. (184)

This process of mediation, labour and its neutralization, depends on the repression of our knowledge about the sacrifices required by others and otherness to enable our imagined freedom of automobility. Elder Brother's words and actions show how bitumen extraction is a form of torture to the earth, and we need to change how we live if our *logos* is to accord with our *bios*.

The story still leaves me wondering, though, about the exigence it creates, the rhetorical opportunity for change. Grant-Davie argues that exigence is "not so much an external circumstance as a sense of urgency or motivation within rhetors or audiences" (277) and defines "rhetorical exigence" as "some kind of need or problem that can be addressed and solved through rhetorical discourse" (265). He points out that "[Lloyd] Bitzer defines the audience as those who can help resolve that exigence... [and] constraints are 'persons, events, objects, and relations which are parts of the situation because they have the power to constrain decision and action needed to modify the exigence'" (266). I am wondering, then, if the constraints are too great to be solved by any audience through rhetorical discourse, or, if the "rhetorical audience" needs to be created by a non-rhetorical intervention, by a parrhesiastes. Maybe if we recognize the ethos of the status quo as the foolishness of Elder Brother, burning the earth to make money will not seem like something reasonable people could do, and then, starting from a recognition of our own irrationality, we might find an ethical space to engage in a rhetoric that sees coexistence with this earth as its goal. Simpson writes,

> The "transfiguration" and "conversion" spoken of by Foucault as indicative of askesis [the practice of severe self-discipline], and, by association, *parrhesia*, is only possible if the truths given in *parrhesia* are both diagnostic and fictive in nature. The truth not only enables one to live in harmony with present reality, but also "enables the subject not only to act as he ought, but also to be as he ought to be and wishes to be." In this way, the truths evoked in parrhesia can carry the governmental functions accorded to them by Foucault and others: one is governed by both reality as it is given and as it will be in the future. (11)

Thus, "fact-based" claims depend, for their persuasiveness, on an explicit or implicit expectation that those "facts" will continue to be operative in the future. Simpson continues,

> Parrhesia seemingly demands a continuous conversion of the self in which the multiplicity of relations between the self and world are altered

> in order to align with the truths and fictions one announces. As both a speech act and a relationship to truth/fiction, parrhesia foments the creative and strategic space in which one can develop the practices necessary to bring about the effects of truth. (11)

The figure of Elder Brother in "An Athabasca Story," unlike tricksters in some stories who attempt to deceive in order to get what they want, "says everything" that he is thinking and, in so doing, puts himself in danger. Even though his words and actions are not perfectly congruent—making him a failed parrhesiastes, at least in the ideal sense that Socrates is a parrhesiastic figure—his actions may lead readers to reflect on their lives and the harmony, or lack thereof, between their lives and their beliefs, their actions and their words.

While the fictive truth revealed through the story of Elder Brother in the bitumen mine may help to re-politicize the ideology of liberal capitalism, the facts deployed in the scene imagined in "From the Bottom of the Pit" operate within conventional rhetoric. This speech may persuade some readers, but is likely to appeal mainly to those who already share the speaker's views. As such, despite its intentions, it may simply add to the cynicism and isolation of the times. "From the Bottom of the Pit" calls on its readers to live differently, but we do not know if the speaker's actions accord with his words since we do not see him act. Therefore, his persuasiveness depends entirely on rhetoric. The final image, of traffic congestion, may point beyond this rhetoric if it enables readers to connect the consequences of bitumen extraction described by the speaker with their own simulacral freedom in automobility. Reading "From the Bottom of the Pit" alongside "An Athabasca Story" makes this connection stronger for me as they point to the more than human relationships that are at stake in extracting and transporting bitumen.

Socrates, as parrhesiastes, acts in accordance with his words to the point of his death, and that death stands as a reminder to subsequent generations to live differently so that other innocents are not put to death. Elder Brother—though he does not die a noble death, does not die at all in fact—suffers, and implicates us in his suffering. He thereby impels us to live differently if we are to avoid his fate, and, simultaneously, impels us to help him, and our other relations, to avoid that fate.

Pipeline Facts, **Poetic** Counterfacts

Metaphor and Self-Deception in a Bitumen Nation

THIS CHAPTER'S MAIN FOCUS is on Enbridge's Northern Gateway Project proposal and two poetry collections created in direct opposition to that project.[1] In some ways the issues involved in Northern Gateway are similar to other pipeline proposals—Keystone XL, Energy East, Line 9 reversal, Trans Mountain expansion. In other ways, each project deals with its own specifics. This chapter does not address the growing transport of bitumen by rail and the threats that poses, threats made especially apparent in the Lac Mégantic tragedy of July 6, 2013.[2] This chapter considers the ways in which individual and cultural self-deceptions are at work in justifying increased bitumen extraction and transportation infrastructure. It examines some ways in which this industry and its associated projects are represented in the print news media, in a Northern Gateway advertisement, by the National Energy Board's Joint Review Panel report, and in the poetry collections. It continues the discussions of *parrhesia* and *ethos* started in the previous chapter and considers both the extent to which poets can serve as *parrhesiastes* in today's society and, given that function, if poetry can counter the self-deception that enables bitumen extraction within the time frame required by global climate change. If Elder Brother implicates us in his suffering and impels us to help our other relations avoid a similar fate, then we need to think through how transporting bitumen also involves transporting the sacrifices involved in its extraction to other locations.

Like Elder Brother, many of the speakers in the poems in *The Enpipe Line: 70,000 km of Poetry Written in Resistance to the Northern Gateway Pipelines Proposal* and in *Poems for an Oil-Free Coast* work to reveal the assumptions underlying arguments in favour of building the Enbridge Northern Gateway Project. By contrast, four newspaper articles show the government's faith in scientific management as the solution to the challenges posed by bitumen development. Three of the four articles present multiple viewpoints as a way of allowing "everything" to be said, but in lacking the strong ethos of the parrhesiastic speaker who risks something in speaking, cannot be read as parrhesia. The poetry collections also present multiple voices and are framed in terms of a duty to speak out (and one of them—*The Enpipe Line*—is framed in terms of danger to the speaker). The Joint Review Panel (JRP) hearings also suggest an opportunity for "everything to be said" but the panel's report makes clear that, as with Alberta's 2006–2007 oil sands consultations, the terms of reference explicitly exclude certain elements and privilege some types of discourse over others.[3] The Northern Gateway advertisement analyzed here deploys poetic conventions to make a rhetorical point: it attempts to establish the ethos of the company as ecologically responsible through the use of poetic language, but the relationship of the speaker to its subject remains hierarchical.

Poems vs. Pipelines

On July 19, 2012, in an article in the *Edmonton Journal* titled "Ottawa Backs Gateway despite US Salvo at Enbridge," James Keller reports on Federal Environment Minister Peter Kent's response to a US National Transportation Safety Board report on the oil spill in 2010 from Enbridge's Line 6B into Michigan's Talmadge Creek, which subsequently reached the

Kalamazoo River. Kent is quoted as saying that the pipeline involved in the spill "is an older pipeline; it is a different set of geographic and technological realities from some of the new major projects being proposed" (A2).[4] The implication of this statement seems to be that the risks of new pipeline projects can be managed more effectively than the risks from pipelines built in the past. While this may not be particularly reassuring for anyone living near an older pipeline, it also begs the question of why "Enbridge did not fix a defect on the pipeline when it was discovered five years earlier" (Keller A2) or how, specifically, "better practices, better technology, better supervision, better regulation and better inspection" (Keller A2) will manage such human errors in the future. On this point, the JRP states that "pipeline technologies, materials, codes, and regulations were developed as a result of lessons learned from previous failures, and scientific research continues to find ways to improve pipeline performance" (National Energy Board [NEB] 43). The JRP report notes, "Northern Gateway said the company and shipping industry are well aware of this factor [human error] and would continue to improve management systems, training, and technology to reduce the risks" (59). Continuously reducing the risks and learning from mistakes suggests a positive progressive narrative for such projects. However, while such a narrative works to incorporate problems, errors, and failures into the idea of progress, it also justifies the sacrifice of certain forms of otherness in the name of that progress (as discussed in Chapter 2). In the case of Northern Gateway, there are many ecological specificities at risk and some, especially caribou and grizzly populations (as discussed below), whose sacrifice is deemed justified by the idea of progress that the pipeline represents.

The slim chapbook *Poems for an Oil-Free Coast* is one attempt to counter the necessity of pipeline building. Offered in a special edition of fifty numbered copies and a regular edition of fifty numbered copies, with proceeds going to the Raincoast Conservation Foundation, it contains "poetry and prose from one Albertan, and eight British Columbian poets" (Red Tower Bookworks). These writers "illustrate the ecological diversity of the north and central coast of British Columbia, now under threat from plans to ship tar sands bitumen from Kitimat" (Red Tower Bookworks). The book arises as a companion to the collection *Canada's Raincoast at*

Risk: Art for an Oil-Free Coast, where these texts also appear. While only sixteen of the fifty pages of *Poems for an Oil-Free Coast* are printed in the regular edition (twenty-four of fifty pages in the special edition are printed), the effect of the book is striking. Each poem is separated by a translucent interleaf sheet of blank Kozo paper, through which it is just possible to read the text, letterpress printed on the subsequent sheet of Zerkall Ingres 90-lb. mould-made paper. Placing this barrier between the reader and the poems, especially given the subject matter of the book—the ecological specificity threatened by Enbridge's proposed pipelines—reminds the reader both of the book's status as go-between, carrying the poet's perceptions of the natural world to the reader and, thus, mediating between the reader and nature, but also of the threat posed to nature by turning it into a conduit for carrying bitumen to the coast—the threat of further separation from and loss of otherness. This threat is mentioned explicitly only in the title of the collection and a brief note preceding the texts. This threat is embodied even more materially in the special edition of the book, which uses, for its cover, yellow cedar boards from the Great Bear Rainforest—one of the ecological areas threatened by the planned movement of supertankers off the BC coast. The prose and poems themselves were not written in response to Enbridge's proposal, but have been taken from their original contexts, reprinted, and made to speak again in this politicized context.[5]

The efficacy of engaging large political questions through poetry—and, aside from the excerpt from Eden Robinson's *The Sasquatch at Home* that opens the collection, all of the pieces are poems—but perhaps especially poetry published in a limited edition, is always open to question. When faced with multibillion-dollar projects backed by transnational corporations, what can poetry do? Indeed, this is a question worth asking of literature more broadly, and which *Unsustainable Oil* tries to engage throughout. In short, literature can create the "pause for thought" (Wong, "J28") in which individuals can question the self-deception into which we are lulled by the cultural hegemony.

Considering the role of poetry in today's society, Alberta poet Alice Major has compared the political context in which Pablo Neruda wrote with that of Geoffrey Chaucer and contrasts both with contemporary Alberta:

> The governments faced by Neruda saw art as dangerous to them. The kings and courtiers served by Chaucer didn't pay him that compliment, but they took his work seriously. To be cultured was a good thing—an entertainment in the original French sense of *entretenir* (to hold something in common, to share). In Alberta, artists are not important enough to be dangerous; if we are to be entertaining, it is in the debased sense of pastime, amusement. In the eyes of the government that has prevailed for the last decade or two, art is just one of the range of options that citizens can select from to pass their time—no more or less valuable than a night at the casino. (18)

I would go further to suggest that the government likely sees a night at the casino as more valuable, since they use the proceeds of gambling to fund social and cultural programs: casinos "make" money; poetry "costs" money. Regardless, Major goes on to state, "it seems virtually impossible that a piece of art—a poem, a painting, a play—will galvanize political change. It is all too possible that we will write only personal lyrics—lovely, intelligent, experimental ones, no doubt—and paint landscapes and abstracts of the same description. That our playwrights and theatres will attract audiences only to comedies, because they are entertaining" (18). That is, artists cannot rise to the level of parrhesiastes, at least in their public function, or, if they do, no one will listen. Major argues further that artists need to be involved in the political process so that "we could live in a province where political leaders feel the creation of music or artwork is something to be actively fostered, that plays are something to go to voluntarily, that citizens as a whole are better off if there is lively production of artwork here. In short, leaders who personally feel that art isn't merely a potential consumable, but what we share, that which holds us together" (18). While I agree with Major's thesis, the incentive of government is in the opposite direction: as the texts examined in *Unsustainable Oil* show, the complicated nature of our relationship to bitumen—a relationship that is symbolic of the culture we inhabit—does not square with the political expedient of simplifying that relationship, of focusing on its benefits and claiming power to manage and mitigate its risks. This focus requires marginalizing art and artists. Rather than art being what holds us together, we are fed a narrative of bitumen as nation-building substance (as described by Rex Murphy below).[6] In this view, transporting bitumen is about transporting economic benefits to other parts of the country. This is, in the end, the conclusion of the Joint Review Panel as well. They note, at the beginning, "Our role was to conduct an independent, science-based, open, and respectful hearing process" (NEB 8) and that "Our recommendations are based on technical and scientific analysis rather than on the number of participants sharing common views either for or against the project" (14). This is particularly interesting given that the panel "received more than 9,000 letters of comment regarding the application. Most of the letters argued against approving the project" (14). In the end, their recommendation to approve Northern Gateway is based on the view that "the project would be in the public interest...[and] that the project's potential benefits for Canada and Canadians outweigh the potential burdens and risks" (71). The decision is one of administrative rationalism rather than democratic pragmatism (Dryzek). And it is a "technical" and "scientific" decision rather than a cultural one, as if technology and science are acultural. The role of literature, or art more broadly, in this process can be to remind us of the cultural nature of science.

Following Dianne Chisholm, Matthew Zantingh argues for the value of poetry as a response to environmental damage:

> Facts and statistics themselves do not lend themselves to far-reaching conclusions unless one possesses the requisite technical or scientific knowledge to fully understand their importance and implications. In a way, our failure to respond to the data of environmental damage is itself a failure of imagination, opening up a space for poetry, and imaginative literature more generally, to act as a crucial component of ecocritical praxis. (626)

Timothy Clark writes, "For Latour the main force of radical environmentalism is that, in openly destabilizing the fact-value distinction upon which so much modern thinking and practice is based, it also demystifies 'Science' as a political ideology, calling scientists to new forms of thought and responsibility" (151). The role of literature, similarly, can be to help readers out of the trance of scientific self-deception into which we can easily fall, a trance that is easily perpetuated when we leave decisions to "the experts," when we view scientific knowledge as separate from cultural knowledge, as non-ideological.

In the Foreword to another collection of poetry that works in opposition to Enbridge's Northern Gateway Project, *The Enpipe Line,* Rex Weyler addresses the question of poetry's political efficacy and concludes that "historically, poetry always wins. Basho lives long after the war lords are vanquished. Rumi and Hafiz survive the Persian empires. Harriet Beecher Stowe outlives American slavery" (11). This, it seems to me, is quite different than suggesting that poems can create or maintain an oil-free coast. The words of Socrates can still speak truth today, but he was forced to drink hemlock in Athens. The poems discussed in this chapter may outlive the petroleum age, but will that age end before pipelines cross salmon streams and oil tankers dock in Kitimat? Patricia Yaeger has wondered if Latour is just talking to academics or if his move to "matters of concern," assemblages, and networks could engage the masses ("The Pipeline"). To put this again in rhetorical terms, is there an exigence here? Will following the poets change the status quo (Latour, "Style" 23)? As Elizabeth Wardle argues, following Pierre Bourdieu, "a person can understand clearly how to speak in ways that are acceptable in particular circumstances, but if not endowed with some recognized institutional authority, all the relevant and appropriate words in the world will not command [those circumstances]: 'authority comes to language from outside, [Bourdieu writes]...Language at most represents this authority, manifests and symbolizes it.'" The small press limited edition publication may have a major cultural impact at some unknown future date, may be, as Lawrence Rainey has argued in the modernist context, "a rarity capable of sustaining investment value" (39), may build authority and cultural capital over time, but this amounts to the same gamble as industry and government make on their ability to create a (e)utopian future: poetry will survive after the petro-economy, on the one hand; the petro-economy will build the future society of free and equal people, on the other. While one may hope that reading

The banks are littered with salmon
dragged from spawning beds,
eye sockets hollowed by ravens,
heads opened, brains licked out
by the bears' hot tongues.

in Alison Watt's "Creekwalker," or

Flukes tapping, feeling the water,
sensitive fingertips of the blind,
he lusts after no one, sings
love songs for light
lyrics for deepest soundings
pastorals for tropical calms.

in J. Iribarne's "Cetology" might create a ripple that energizes the movement to stop bitumen extraction and pipeline building, the likelihood of that seems slim, as Major argues. The likelihood of this limited edition chapbook acquiring the necessary institutional authority to be accepted by a powerful discourse community, to alter the circumstances into which it speaks, also seems slim. The Joint Review Panel of the National Energy Board, by contrast, is authorized to speak by a powerful institutional context immediately; a great deal of effort would be required by many other speakers with authority in other contexts to erode the panel's authority in the short term. Furthermore, Northern Gateway is able to capitalize on the authority of the JRP by deploying its discourse in other contexts, such as on its website where it declares the company is "Working hard towards meeting the 209 conditions set out by the Joint Review Panel" (Northern Gateway). Thus, even though thousands of individuals spoke in opposition to the pipelines, the Joint Review Panel (like the Oil Sands Consultation Multistakeholder Committee before it) served as a medium for managing that opposition and converting it into a positive recommendation with conditions attached. In order to be successful, opponents will need to speak in a context in which they have institutional authority.[7]

Poems vs. Advertising

In the Introduction to *The Enpipe Line*, Christine Leclerc writes of participating in a Greenpeace-organized protest of Enbridge's proposed Northern Gateway Project by chaining herself, along with other activists, to a door in One Bentall Centre, Northern Gateway's corporate offices. It was after the chains were cut by police that "the image of a poetry-jammed pipeline struck" her (16). Soon after, during an "artist's residency with the Gabriola Institute of Contemporary Art" (16), Leclerc conceived of "The Enpipe Line as a 1,173 kilometre-long poetry collaboration, designed to go dream vs. dream with Enbridge's pipeline proposal" (16). The project has now grown to more than seventy thousand kilometres in length, when, as I understand it, each centimetre of text is printed at a scale of one kilometre long. The text is also supposed to be printed one kilometre high to match the right-of-way of the pipeline, which is an interesting counterfactual proposition given that in reality the text is printed in a standard font. Some poems in the collection address the Northern Gateway Project specifically, while others discuss opposition to similar projects. The neologism "enpipe" is defined as "To block up and/or fill a pipe to bursting" (15). Each poem's title is followed by a kilometre mark.

Thoughts of an Anthropologist (178.8 km) by Kaitlin Almack

While we can separate self
By skin
The person is not divisible
The world in our body
Is never escapable

Our bones stacked like forest leaves
Burnt as sticks on the water
Flesh dripping as candle wax
Into the glacial river

You go into the forest
Caribou
Birth order and age
To be moral is to be
related

Our brothers bodies
Are blackbirds
That come to the house
on the busy street

Our home is a town
We have never been
but the spirits
Are kept there

It is difficult to consider these poems in terms of parrhesia because it is impossible to know whether there is congruence between the poets' words and actions. Indeed, a simplistic caricature of the necessity for such congruence is frequently used to dismiss opposition to bitumen extraction. When Robert Redford spoke out against the Keystone XL Pipeline and bitumen extraction in September of 2013, Alberta Premier Alison Redford was quick to dismiss his concerns: "I've really got to question how people who are using energy flying on planes can make these sorts of comments and assume they're going to have any credibility" (qtd. in Sinnema A11).[8] However, these poems are clearly not the "comic poets" who were marginalized in ancient Greece. As I've been arguing throughout *Unsustainable Oil*, one of the values of fictional literature is to promote dialogical engagement with the reader/listener, unlike the frequently monological narrative of the current hegemony. These poems point to the things that hold us together. To repeat Latour's inversion of Archimedes: "Give me one matter of concern and I will show you the whole earth and heavens that have to be gathered to hold it firmly in place" ("Critique" 206). Almack's anthropologist notes, for example, in contrast to the traditional anthropocentrism of that discipline, that "The world in our body / Is never escapable" and

"To be moral is to be / related." Such recognition of relationship to and dependence on otherness suggests a broader range of concerns than the "balance" sought by pipeline proponents or the Joint Review Panel. We may, then, try to think of these poems in terms of a collective parrhesia, a call for a congruence between our collective words and our collective actions.

Say (246 km) by Melissa Sawatsky

this line is full of words
that burst without warning, collateral
damage of constructed intention, which is to say,

there are things you can claim. All the way
from Bruderheim to Kitimat, you can megaphone
phrases—safe passage,
62,700 years of person-employment,
maximum environmental protection,
open and extensive public reviews—
intangible words in corporeal space.
Which is to say,
there is nothing you can say. The land
doesn't want your pipes for veins,
doesn't need the petroleum/condensate
exchange. Though at the centre of the argument,
the land is not the victim. It has spent eons learning
how to survive every age-old oppressor.

In Sawatsky's poem we see an example of the repurposing of industry discourse—"safe passage / 62,700 years of person-employment" etc.[9]—which is countered with a gesture toward silence—"there is nothing you can say." While this might be read as equivalent to the dismissive rhetorical strategy of industry and government—the opposition doesn't know what they're talking about, we need science and facts—it is also a gesture to an alternative context, an attempt to speak for the land, to add a missing voice to the conversation. Interestingly, this strategy has also been incorporated into Enbridge's advertising.

In a series of two-page advertisements, which I encountered in *The Walrus*, readers are presented, on the verso, with an image of a kelp forest viewed from below with sunlight streaming through the water, and, on the recto, with a short poem followed by a prose passage about the responsible construction of the pipeline. There is a great deal of white space on both pages, perhaps meant to be suggestive of the small environmental footprint of the project. In one such ad, we read:

> **The ocean—**
> Vast. Deep.
> A limitless pool of life.
> A playground for the tiny and
> giant things that live within it.
> And a gateway to the other side.
> The ocean should remain an ocean.
> Always.

This ad reverses the typical industry strategy of starting with facts and then linking them to a counterfactual narrative of future progress. Here, the poem gestures toward the immensity of the ocean, its "limitless... life" seemingly unthreatened by human activities, before it introduces the ocean's specifically human value as "gateway to the other side." How these two values coexist is unexamined in the poem. Instead, we have the juxtaposition of two metonymies—the ocean is renamed as both "a playground" and "a gateway"—and an implied relationship between these realities: at the very least, it can be both of these things at the same time. Further, we are told that the ocean should remain this way "Always."

The poem is followed by a short paragraph, in slightly lighter ink, appearing almost as a reflection of the poem, which suggests how we can maintain the ocean as it is:

> The Northern Gateway Pipeline will
> protect our oceans by ensuring all
> tankers are guided by certified BC
> Coast Pilots with expert knowledge of
> BC's coastline. Because a better pipeline
> will not be built at the expense of
> making other things worse.

Here we have the standard industry appeal to "expert knowledge" but also an interesting transference of agency to "The Northern Gateway Pipeline."[10] It is the pipeline itself that will "protect our oceans." When these two texts are read together we have the ocean's metonymic change of name into positive anthropocentric terms—playground, gateway—and the declaration of protection for these things by the pipeline itself. The possessive pronoun "our" suggests human ownership of the oceans, which we, paternalistically, preserve as a "playground" for the "things" that live within it. A playground is, most obviously, to me at least, an outdoor area provided for children to play safely under adult supervision: conceiving of the "limitless" ocean in this way is quite different than a view of nature that "has spent eons learning / how to survive every age-old oppressor" or a relationship with nature in which "our brothers bodies / are blackbirds." Rather than a relationship of brotherhood, of equality, or even submission to non-human creatures, the ad conveys a proprietary and controlling view of nature. The ad concludes near the bottom of the page with "Find out more at gatewayfacts.ca" and the Northern Gateway's logo. The phrase "find out more" implies that we have found out something from this advertisement, and, while we may have, what we have found out are not "facts." The only fact in the ad is that tankers will be "guided by certified BC / Coast Pilots." The rest of the ad is a series of upward counterfactuals. The concluding "Find out more at gatewayfacts.ca" suggests that the counterfactual vision of the ad is based in "facts" that readers can "find out" on the website.

Poems vs. Profits

The Enpipe Line was created in explicit opposition to the status quo and promotes what Joe Oliver would consider, as discussed in the Chapter 3, a "radical environmental agenda." Although some of the poems engage with the language of science, they do not rely on the rhetoric of science or the knowability of a closed economy (as discussed in Chapter 1)—which, I feel, opposition groups too often slip in to—for whatever persuasive force they have.

Although there are some wonderful poems in the collection, I am left uncertain as to the exigence of such interventions.[11] Can poetry stop a pipeline as the image of the enpiped pipeline suggests? Can it do it, not in the eternal timeline of truth that Weyler refers to or the geologic timeline that Sawatsky gestures toward at the end of her poem, but *now*, in this time of tough oil and capitalist profit-taking?

Pipeline (19.44 km) by Kathryn Mockler

The

ones

who

are

not

there

don't

have

to

think

about

it.

Perhaps in some post-oil future these poems will remind humans of what was lost and persuade them to avoid making a similar mistake in the pursuit of another form of progress. I fear that, in the mean time, for most people most of the time, Mockler is right: we don't have to think about it and we prefer not to because we "we know enough to want not to know" (Stoekl, "Unconventional Oil" 40).

Science, Self-Deception, and the Location of Hope

While environmentalists target counterdiscursive rhetoric at government and industry propaganda, bitumen continues to be extracted because the public is deceiving itself about the impossible choices involved in that extraction. In *Fooling Ourselves: Self-Deception in Politics, Religion, and Terrorism*, Harry C. Triandis states that those who engage the services of "astrologers, fortune-tellers, palmists, readers of tarot cards or crystal balls, mediums, prophets of all kinds, faith healers, psychic surgeons, psychics of all kinds, parapsychologists, and other makers of miracles... pay for the hope" (1) these charlatans offer. Further, "in these self-deceptions, individuals believe that they have some control over events, that is, they can predict the future or they can have good luck. They ignore any evidence that is contrary to their assumptions" (3). Now, I don't want to equate science with the pseudo-sciences that Triandis is talking about, but I think there is a parallel when science is mobilized to predict the future, when scientific facts are linked to rhetorical counterfacts. That is, science can be used to offer hope much like that offered by a fortune-teller. Scientific monitoring can tell us something of the consequences of past activities—we can measure particulates that we have already released into the air—but this is different from predicting what will happen in the future (a difference climate change deniers have capitalized on effectively, as Latour shows). Science, rather than looking backwards, as with evolutionary biology, to explain how we have arrived at the current situation, is projected into the future to offer hope (or, in a minority strain, doom at a coming eco-apocalypse) that the status quo of resource extraction, and the comforts of energy and jobs that it provides, can continue into the foreseeable future.[12] The Joint Review Panel, for instance, notes that, "Some environmental burdens may not be fully mitigated in spite of reasonable best efforts and techniques. Continued monitoring, research, and adaptive management of these issues may lead to improved mitigation and further reduction of adverse effects" (NEB 71). Despite the sacrifices required by building these pipelines, it might be possible to create ways to reduce or compensate for those sacrifices in the future; this is projecting faith in science into the future.

Discourse, in the words of John Dryzek, is "a shared way of apprehending the world...[that] enables those who subscribe to it to interpret bits of information and put them together into coherent stories or accounts" (8). However, though Dryzek recognizes that "the way a discourse views the world is not always easily comprehended by those who subscribe to other discourses," he suggests that "complete rupture or discontinuity across discourses is rare, such that interchange across discourse boundaries can occur, however difficult it may sometimes prove" (8). The poems in both collections attempt to push against the rhetoric of pipeline/nation building by pointing to the many things this rhetoric leaves out and imagining a range of counterfactual possibilities resulting from proceeding with those developments. The question of exigence, again, asks whether there are possibilities for interchange across the boundaries between the Joint Review Panel and *The Enpipe Line* or between Peter Kent and *Poems for an Oil-Free Coast*?

We might think about this division between discourses as worldviews in relation to Michel de Certeau writing in *The Practice of Everyday Life* about a story told by Immanuel Kant:

> Where I come from, [Kant] writes, "the ordinary man" says that charlatans and magicians depend on knowledge (you can do it if you know the trick), whereas tightrope dancers depend on an art. Dancing on a tightrope requires that one maintain an equilibrium from one moment to the next by recreating it at every step by means of new adjustments; it requires one to maintain a balance that is never permanently acquired; constant readjustment renews the balance while giving the impression of "keeping" it. The art of operating is thus admirably defined, all the more so because in fact the practitioner himself is part of the equilibrium that he modifies without compromising it. (73)

The charlatans or magicians are the bullshitters who seek to trick the public with their fancy rhetoric into accepting their discourse position.[13] The tightrope dancers are the artists who seek to view the subject as a matter of concern by constantly readjusting their positions with each step they take.[14]

The movement between discourses depends on individuals being acted upon by external forces: we are tricked by the magician or moved by the artist to view bitumen differently. How, though, does this process work when we are the ones tricking ourselves about bitumen, when the issue is one of self-deception? It is very difficult to move from a view that accepts the charlatan's tricks to a discourse that sees our every movement as part of a high-wire act, where our own bodies are implicated in the extraction of oil, even if the latter is more true. We know enough about its truth to want not to know: we know that knowing this would require painful changes to our lives.

Managing a Bitumen Nation

In the *National Post* of March 17, 2012, Rex Murphy defends bitumen development against a narrow environmentalism that ignores the human benefits of industry—"Having a job and earning a living is a great thing," Murphy writes (20). He goes on to connect this virtuous aspect of bitumen extraction to the cod fishery collapse of the early 1990s: "Great social misery was averted because of the oil boom...It is a great story of modern confederation: how Alberta, in particular, modified and mitigated the misery of Newfoundland" (20).[15] However, the connection to the cod collapse suggests something else to me: the danger that history is repeating itself due to our ongoing self-deception about our ability to manage nature.

In *Managed Annihilation: An Unnatural History of the Newfoundland Cod Collapse*, Dean Bavington examines "the role that scientific management played in the destruction of the northern cod" (xxv). In short, Bavington shows that management, rather than protecting the environment, or enabling sustainable resource development, *created* the collapse of the cod: "the northern cod was *scientifically managed* out of existence" (2), he writes. Further, "Rather than being a case of ignorance, neglect, or unwise management, prior to the moratorium, the northern cod fishery was presided over by one of the world's most comprehensive renewable resource management systems" (Bavington 2). Gismondi and Davidson make a similar point about the threat such industries pose to labour:

> The experiences of this increasingly migratory staples workforce should set off warning bells for Albertans reliant on tar sands jobs. Unless communities, governments, and unions plan otherwise, when the natural resource is exhausted, corporations do not create new jobs to replace them, and workers generally do not get "retrained." Instead, they become subject to the geographic whims of the global primary industry labor flow. Unfortunately, when the Alberta energy boom subsides, and if no alternate energy sources emerge, many of the employment fixes available to former extractive workers who lost jobs in one region in the 1970s and 2000s may no longer be there, if there is no more energy to fuel another round of extraction. (91)

Albertans may find themselves, or their children or grandchildren, left with no jobs and no land capable of supporting them.

The management of bitumen and its environmental consequences over the past forty years has been critiqued by many, but another piece in the *Edmonton Journal*—"Oilsands Monitoring Program 'on Track'"—suggests a move towards a much more comprehensive resource management system. It quotes Federal Environment Minister Peter Kent as saying "this [monitoring program] is evidence that we can be environmentally rigorous and also support the development of the oilsands" (Klinkenburg, "Rethinking" A7). It is hard to argue that moving away from the former system of ignorance, neglect, and unwise management in the bituminous region of the boreal forest is a bad thing, and having scientific data is part of being environmentally rigorous, but that, in itself, is not enough.[16] As the analogy with the cod fishery shows, one can have a world-class management system in place and still fail. Indeed, it might make environmental failure more likely, if, as with the cod, the data is interpreted within a context of maximizing profit or growth. To return to a point made in the Introduction of this book: numbers can mean very different things depending on the context and the value system in which they are interpreted.

Fred Wrona, senior science advisor for Environment Canada, said of the new monitoring system, "We are using science and gaining knowledge to get a better handle on what may be happening as a result of oilsands development" (Klinkenberg A7).[17] I want to suggest that faith in scientific management easily becomes a form of self-deception that perpetuates the status quo. As Jeff Gailus notes, "a monitoring system is designed to monitor, not regulate" (39), so even though "ten years from now" we will have data on

> how fast caribou populations...have declined, how much toxic pollution has been released into the Athabasca River...and whether technological advancements have reduced overall water use and greenhouse gas emissions...The data itself and the scientists who provide it will not change a thing unless the government changes the rules in order to mitigate or eliminate the nasty consequences of turning bitumen into oil. (40)

Gailus continues, "Incredibly, both the federal and Alberta governments are working on ways to 'streamline' the environmental assessment process so they can start digging and drilling even faster" (40). During the final decades of the PC dynasty in Alberta, it was easy to be cynical and conclude that the incompetence of the Regional Aquatics Monitoring Program was a calculated delay tactic (as Schindler claims in *Tipping Point*), as is the implementation of the current system: we are now in the position of having to catch up on the science that wasn't done in the past and that process will enable the status quo to persist a little longer.[18] In the meantime, the "lack of scientific certainty" continues to work in favour of the status quo, and it remains unclear how the New Democratic government will address this regulatory challenge.

As Jennifer Grant of the Pembina Institute points out about the joint federal-provincial monitoring program, "This is a three-year study program... but applications for development in the region are still being approved. Monitoring is only valuable if it is linked to decision-making" (Klinkenberg A7).[19] More than that, though, monitoring is only one piece of the complex task of decision making. Having more information could help to counter our collective self-deception about the costs of bitumen extraction or it could help to perpetuate it by making us continue to hope that responsible development is possible, by making us continue to deceive ourselves about the consequences of our actions. Having more data might just make it easier for us to go on deceiving ourselves—indeed, I think this is how we should read most industry and government discourse: it tells us what we would like to believe.[20]

The fourth newspaper article, "Caribou at Risk in Oilsands Regions," dated July 7, 2012, provides further evidence of the limits of managerialism in dealing with bitumen. It reports that an assistant deputy minister in Environment Canada's environmental stewardship branch, Colleen Volk, warned minister Kent in a memo of September 28, 2011 that "There is an 'elevated risk' that threatened populations of boreal caribou in Western Canada will disappear before oilsands developers have the chance to restore old-growth forests being disturbed for industrial expansion" (De Souza A13). The article reports that Volk prepared the memo in "the context of a legal battle between environmental groups that have taken the federal government to court for failing to enforce its own endangered species law." However, "Kent decided to dismiss the regional concerns after determining the species was not threatened on the national scale" (De Souza A13). The article concludes by noting that "Kent has said the government is now reviewing how to 'improve' the Species at Risk Act through new legislation he hopes to introduce later this year [2012]. He said the changes could help distinguish between protections for species that are at greater risk versus those that are only at risk in some regions" (De Souza A13). The logic of management and mitigation has, in this context, led to a proposal to fence off an area to preserve a population of boreal caribou until reclamation has reached such a stage that they can again be released (Canadian Press, "Ottawa").[21]

In regard to the management of ecosystems, the Joint Review Panel for Northern Gateway found that,

> even considering Northern Gateway's proposed mitigation measures and our conditions, the project would cause adverse environmental effects, after mitigation, on a number of valued ecosystem components. These include the atmospheric environment, rare plants, rare ecological communities, old-growth forests, soils, wetlands, woodland caribou, grizzly bear, terrestrial birds, amphibians, freshwater fish and fish habitat, surface and groundwater resources, marine mammals, marine fish and

> fish habitat, marine water and sediment quality, marine vegetation, and marine birds. We do not recommend a finding that potential effects, from the project alone, are likely to be significant for any of these valued ecosystem components. (NEB 57)

Further,

> We also considered cumulative effects for each valued ecosystem component. In two cases we recommend that project effects, in combination with effects of past, present, and reasonably foreseeable projects, activities, and actions, be found likely to be significant. These were effects on woodland caribou (for the Little Smoky herd of the boreal population of woodland caribou and the Hart Ranges, Telkwa, Narraway and Quintette herds of the southern mountain population of woodland caribou) and eight grizzly bear populations that would be over the linear density threshold. (57)

However, despite this recognition, "Considering the overall benefits and burdens of the project, we recommend that significant effects in these two cases be found to be justified in the circumstances" (57). It is worth pointing out that the JRP did not consider the upstream nor downstream impacts of the project (17)—potential increases in bitumen extraction projects that may result from the approval of the pipeline, nor increases in greenhouse gas emissions from refining and combusting the petroleum products transported through the pipeline. Therefore, the "cumulative effects" are being defined in quite narrow and specific terms here. The impacts to the woodland caribou populations whose ranges intersect with the pipeline have not been connected to the impacts on the woodland caribou populations in the areas where bitumen extraction is occurring.[22]

Applying science in the service of economic growth takes what should be a set of debatable questions around the relative merits of jobs, taxes, trees, caribou, tradition, modernity, regionalism, globalization, attachment to place, immigration, freedom, inhabitation, in short a complicated balancing act, involving a series of choices, and subsumes them to the idyllic future promised where improved technology enables pipelines to transport bitumen without threat of spills, reclamation restores land to its pre-development state without impacting existing ecology in the meantime, and environmental monitoring leads directly to technologies to mitigate any deleterious substances emitted into the air, water, or soil. While the Joint Review Panel's report points out some of the complexities involved in the decision making, that caribou and grizzly populations will be sacrificed by building Northern Gateway, this complexity is erased in the advertisement: "a better pipeline / will not be built at the expense of / making other things worse." Rather, the reality is that even the best pipeline will make some other things worse. The JRP claims that significant impacts, making things worse, for caribou and grizzly are "justified"; it does not say why. There is evidence, in the submissions to the JRP, to suggest that a majority of people disagree with this assessment. However, we must again consider Burke's point that "a 'good' rhetoric neglected by the press obviously cannot be so 'communicative' as a poor rhetoric backed nation-wide by headlines" (25) and the mainstream press continues to give prominence to the narrative of scientific management.

Transporting Bitumen Discourse: Metaphor and Self-Deception

A danger is that we *will* believe, that once we have enough data, we will be able to manage the situation effectively.[23] The notion that we can analyze a situation or a substance (such as bitumen), use it for our own ends, and then reintegrate it without losing anything of value is a very effective trick, which persuades us that we are not perched on a high wire, that we are not in danger. This trick operates through the logic of metaphor; we have a transfer, a carrying across, of meaning: the substance "bitumen" moves from being a particular hydrocarbon to being a source of freedom; bitumen moves from being the object of scientific inquiry to being the means of securing an ideal future. This transfer enables the status quo of resource extraction to persist, and we need to recognize the operation of hegemony in such metaphors because accepting them, while on a high wire, may cause us to lose our balance. Paradoxically, it is through destabilizing such metaphors, such givens, that we can continue our high-wire walk. Destabilizing hegemony is the function of a parrhesiastes.[24] Thus, the poems in both collections might be viewed in those terms whether or not the speakers in the poems or the authors of the poems literally put themselves at risk through their speaking.

night gift (790 km) by Rita Wong

Water is to land what the voice is to the body.
—Kaluli knowledge, via Steven Feld and Walter Lew

make space and let the night speak through you—what will
the darkness say? will it sigh the song of nightcleaners, the
lament of the wrongly imprisoned, the rage of the ragged, the
dispossessed? how will the night take you back? will you be
the vessel for earth shatter, hydro poison, ancestral revenge?
perhaps steady weeds, growing irrepressibly into the cracks,
urban repurposing, straddling both the drugs that kill and the
ones that heal? the globe moves around the sun, unstoppable,
feeding pine trees and petro-states alike, giving us the days
and nights by which to stand with the trees, what the industry
calls overburden, or to die more rapidly, more stupidly, by
peak oil. as rivers and oceans fill with carcinogenic wastes
from the petroleum-plastic-supply chain, the political
systems that follow, stuffed full of suncorpse and tired old neo
colonial ego that refuses to stop growing until it reaches the
limits of the planet's patience. who knows what alliances and
monkeywrenches will be enough to stop the greed of the
greasy machine? what I do know is that humble migrants
who've traveled the ocean know its wisdom better than an
arrogant elite that doesn't heed the world's necessary stories.
jail the stories and the storytellers, but they will keep speaking
the night, until empire expires, with or without the multitudes
alive. in this race may we be ready to move fast, yet steady
enough to encompass musicians and lake gatherings, forests
and guerilla gardens, fueled by a love more immense than
the unnatural systems we've inherited. we need to live the

> world that is possible even while we struggle through war.
> respect living coasts and fluid watersheds, not murderous
> imperial borders. in grief and in celebration, in fear and in
> courage, in anger and in compassion, the night replenishes us
> so that we may continue to embody her songs.

Reading Wong's poem "sort by day, burn by night" from her collection *forage*, Zantingh argues that the poem "asks us to consider how the material objects of everyday life emerge from networks that connect disparate places and people together" (624). We can see a similar approach at work in "night gift" with its reference to the "rivers and oceans [that] fill with carcinogenic wastes / from the petroleum-plastic-supply chain," a supply chain that flows through our lives and into our bodies. The poem reminds us of the material world sacrificed and the choices readers can make to "stand with the trees" or to "die more rapidly, more stupidly, by / peak oil." Such a choice does not mean that it is possible to get outside of the network of toxicity, as Zantingh notes, but that we can take the "implications of material objects more seriously" (640). The "night gift," which "replenishes us / so that we may continue to embody her songs" may be inadequate, as the poem suggests—by the time the "empire / expires" it may be "with or without the multitudes / alive." Nevertheless, the "planet's patience" is limited and will run out. In the meantime, humans have a role to play; we are called on to "keep speaking / the night." Zantingh claims that "Wong's poetry provides a useful bridge between the two discourses [toxic discourse and material culture] as she negates the anxious negativity of toxic discourse with her poetic optimism in the ability of humanity to reconfigure its relationship to objects" (639). In "night gift" we see Wong's optimism in the "humble migrants"; "stories and storytellers"; "musicians and lake gatherings"; "forests / and guerilla gardens"; and the "love more immense." In opposition, then to the upward counterfactuals of bitumen as metaphor of freedom, Wong offers a view of relationships that exist and persist alongside the toxicity of peak oil and its consumer culture.

Zantingh points out that "consumer culture operates on the general principle of generating desire for further consumption through marketing, in the process driving consumers to purchase more and more with little to no self-reflection" (629). Poetry does not exist outside of this culture, but it can serve to enable self-reflection, it can help us to wake up, to recognize some of the charlatans' tricks, to recognize that we are perched on a high wire and that our capacity to keep our balance is dependent on our connections to all of the others and other others who are also trying to keep their balance on their own high wires. The Joint Review Panel quotes the definition of *environment* from the Canadian Environmental Assessment Act, 2012 as including "(a) land, water, and air, including all layers of the atmosphere; (b) all organic and inorganic matter and living organisms, and (c) the interacting natural systems that include these components" (NEB 45) and notes that "the interacting natural systems, or ecosystems, are all connected and depend on one another" (45), but, ultimately, the JRP's recommendations operate within a view of the environment as something that humans can manage, as something containing "components" that can be added or subtracted, rather than as the product of the relationships among those components. Poetry can help us see some of the limits to such a hegemonic managerialism.

The discourse surrounding the Northern Gateway Project shows that as much as there are practical challenges in transporting bitumen from Northern Alberta to various world markets, these challenges are engaged in discourse, by transporting a set of symbolic associations from one location to another, for one side, and by disputing those associations, for the other. Is bitumen freedom or is it toxicity? Is the ocean a playground or is it a home? While the practical challenges can be met with scientific and technical knowledge (it is possible to build pipelines), the symbolic challenges are more intractable. If pipelines are built, critics will still argue that bitumen is dangerous and toxic; if pipelines are not built, proponents will still argue that bitumen is the means to a (e)utopian future. Whether or not any pipeline is built, whether or not many pipelines are built, we need to look beyond scientific and technological management in thinking about our relationship with the future. What is the relationship between our actions toward bitumen today and the future those actions are creating?

Reading poetry, reading literature, can help us pause and think through those questions, can interrupt the smooth functioning of a system that claims scientific knowledge in the service of maximizing profit, can wake us from the dream of limitless growth.

Oil Desires

Appetites and Fast Violence in the Bituminous Sands

IF THE HOPE OF TECHNOLOGICAL LIBERAL MODERNITY is embodied in the idea that "petroleum products from places like the Athabasca tar sands might not only run our cars and surface our highways, but *feed us* as well" (Kroetsch, *Alberta* 274), then, following the adage "you are what you eat," we are in the process of becoming oil. Andrew Nikiforuk quotes US sociologist Fred Cottrell on the fact that a petroleum-based food production system "consume[s] more energy than it produce[s]" (*Slaves* 84). Noting the end of the "end of history" thesis, the thesis that imagined the worldwide society of free and equal people and an end to human suffering and hunger, Allan Stoekl points out,

> The great myth that Man "forms himself" by forming, and transforming, brute matter is over. The idea that Nature is dead is over because fossil fuels were not made by Man, they were only extracted by "him." They are brutally natural, and their shortage too is a natural shortage (their *lack*

> is natural). And when a profound, irremediable shortage of those fuels supervenes, history opens back up. History will not, as some critics of the "end of history" thesis claimed, return merely as localized struggles and revolts that put the superpowers on the spot. Instead, History now is the fight for a resource that will allow History as we have come to think of it—the flourishing of civilization and the establishment of the definitive dignity of Man—to continue and triumph. No one yet wants to think about how History should continue in the absence of an adequate supply of fossil fuels. It is too horrible to think about. Human die-off is quite natural, but it also constitutes an incontrovertible historical event. With the finitude of cheap energy, alas, the end of history is itself finite. (*Bataille's Peak* x)

The end of history is finite because it means the end to the surplus of cheap energy that enabled cheap food. Perhaps, then, the idea in Richard Van Camp's *Godless But Loyal to Heaven* that some people will in the future be raising other people as cattle (15) to feed to the shark throats, a new zombie race, should not be shocking. But it is.

This chapter will connect *Godless But Loyal to Heaven* and *5000 Dead Ducks: A Novel about Lust and Revolution in the Oilsands* through a focus on the appetitive. It will argue that the promise offered by hydrocarbons is a false one—the overcoming of human slavery and the satisfying of human desires that it promised is illusory. Energy slaves also require human slaves.[1]

Discussing the prevalence of consumption within late twentieth- and early twenty-first-century capitalism, Jean Comaroff and John Comaroff write,

> The workplace and honest labor, especially work-and-place securely rooted in local community, are no longer prime sites for the creation of value. On the contrary, the factory and workshop, far from secure centers of fabrication and family income, are increasingly experienced by virtue of their closure: either by their removal to somewhere else—where labor is cheaper, less assertive, less taxed, more feminized, less protected by states and unions—or by their replacement by nonhuman means of

> manufacture. Which, in turn, has left behind, forever [sic] more people, a legacy of part-time piecework, menial make-work, relatively insecure, gainless occupation. For many populations, in the upshot, production appears to have been replaced, as the *fons et origo* of capital, by the provision of services and the capacity to control space, time, and the flow of money. In short, by the market and speculation. (781)

This shift away from industrial capital to the financialization of economies is also coupled, though, as Nicole Shukin points out and as I mention in Chapter 2, with the intensification of extractive resource capitalism. Capital still needs something, material, to speculate about—oil prices or housing starts or tech start-ups—and to power its "virtual platforms" (despite the "immateriality" of the Internet, it depends not only on software code and the labour of programmers but also on servers, cables, devices, electricity, and its associated infrastructure,[2] and then there is the question, obscured by consumer culture, of what becomes of the hardware once it is obsolete) (see Zantingh, especially 622–25).

Bitumen as Fast Violence

The first story in Van Camp's collection, "On the Wings of This Prayer," details a prayer thrown from the future to two people in the present, warning that continued bitumen extraction will wake a Wheetago who is buried in the sands. In "The Fleshing," we learn that the people who received the prayer are Bear, the young Tlicho warrior-in-training, and Snowbird, the blind medicine man. The two stories are linked by the titular prayer from the first story. The prayer is thrown from the future in order to persuade the receiver, and by extension the reader, to "wake up" (Van Camp 17) and stop the tar sands, because otherwise a Wheetago will rise and infect others, creating a new race—the Shark Throats, Hair Eaters, Boiled Faces—a race of powerful zombies (Van Camp 10) who will consume all animal life: "If you are reading this, please know that I tell you these things because I love you and wish for the world a better way. I have sent this back to tell you this, my ancestor: the Tar Sands are ecocide. They will bring Her back. In both stories, it is the Tar Sands to blame. This is how the Wheetago will return" (Van Camp 12). Nikiforuk describes bitumen as "the product of ancient marine life (largely sun-baked algae and plankton). Some 200 million to 300 million years ago, geological forces started to compress and cook the dead plants and creatures, then degraded the remaining mess with bacterial activity. Good cooking results in light oil. Bad cooking makes bitumen" (*Tar Sands* 12). Or zombies? If the end of Elder Brother in "An Athabasca Story" is the result of a cultural repression of our awareness of dependence on our relations, the zombies in Van Camp's fiction can be read as the return of that repressed. The extraction of energy from dead plant and animal matter animates our lives, but when that dead matter is itself animated, takes on a life of its own, it threatens to consume us. Allegorically, we can read the Wheetago as turning the "slow violence" (Nixon) of pollution into fast violence; turning the "slow death" (Berlant) of ecocide into the fast death of zombie apocalypse.

The Comaroffs describe "millennial capitalism" as "capitalism invested with salvific force; with intense faith in its capacity, if rightly harnessed, wholly to transform the universe of the marginalized and disempowered...It accords the market itself an almost mystical capacity to produce and deliver cash and commodities" (785). Nikiforuk has dubbed this "the economist's delusion," explaining, "In insisting that labor, markets, and technology make the world go round, neoclassical economists have ignored the primary source of all wealth: energy" (*Energy* 131). It is not possible to create surplus value from nothing. In convincing ourselves that we can, we become increasingly indebted. Nikiforuk quotes financial analyst and anthropologist David Graeber as saying,

> You have a population all of whom are in debt, and who are essentially renting themselves to employers to do jobs that they almost certainly wouldn't want to do otherwise, to be able to pay those debts. If Aristotle were magically transported to the US he would conclude that most of the American population is enslaved, because for him the distinction between selling yourself and renting yourself is at best a legalism...We've managed to take a situation which most people in the ancient world would have recognized as a form of slavery and turned it into the definition of freedom. (*Energy* 72)

The Comaroffs declare, "The globalization of the division of labor reduces workers everywhere to the lowest common denominator, to a disposable cost, compelling them to compete with sweatshop and family manufacture. It has also put such a distance between sites of production and consumption that their relationship becomes all but unfathomable, save in fantasy" (784). Terre Ryan makes the point that "the bodies of workers are managed like other resources." Stoekl follows Heidegger in this regard, noting that

> our being, our subjectivity, is a quantifiable term that is a function of the very same movement, the very same bringing forth as *techné*, that renders the world a quantifiable mass ripe for exploitation. And such a subject, immediately transformed into an object, a standing reserve, warehoused in an institution (concentration camp, prison, army, hospital, school, freeway, suburb), is itself ripe for use and disposal. (*Bataille's Peak* 132)

It is in this sense that Van Camp's description of humans raised for slaughter is, not unlike Jonathan Swift's "A Modest Proposal," the logical conclusion of capitalist practice: "They spoke of lakes now, filled with humans swimming in their own blood...They just keep adding more and more people to these lakes. There were lines of people for miles, as far as we could see" (15). People are resources like any other. Shukin has considered "resource extraction as pathogenic activity" and argues, "Once a material is coded as a resource, it can only be valuable once it has been extracted" ("Place"). The Boiled Faces, if they are capable of thought, think of all flesh in this way.

In discussing the rise of zombies as an active force in contemporary South Africa, the Comaroffs describe their intention to interrogate "the fact that [neoliberal capitalism] appears to offer up vast, almost instantaneous riches to those who control its technologies, and, simultaneously, to threaten the very livelihood of those who do not" (782). In Alberta, the narrative of bitumen development has promised every citizen a share of the riches,[3] but, in practice, as *Unsustainable Oil* has argued throughout, many individuals and groups are excluded from sharing in the riches and the market does not mystically create those riches out of thin air. Rather, they are created through processes requiring high-energy inputs, dangerous labour conditions, and environmental degradation.

In both *Godless But Loyal to Heaven* and *5000 Dead Ducks*, irony is used to show that it is not just peoples' livelihoods that are threatened, but also their lives. Indeed, in South Africa, the Comaroffs show, those who are controlled by such technologies of neoliberal capitalism are thought to be zombies, the dead reanimated, and those who are deemed to benefit from the labour of such unnatural creatures are in danger of violent retribution: "the constantly reiterated suspicion, embodied in the zombie, that it is only by magical means, by consuming others, that people may enrich themselves in these perplexing times" (792). Ultimately, the lives of those who (think they) control the technologies of capitalism are as much at risk as those who are controlled. These literary texts can help us to see, in a new way, the myth of liberal capitalism that we are living within; they can help us to wake up, to see that even those of us who may think we are benefitting from extraction capitalism are threatened by it.[4]

At the beginning of "The Fleshing," a pregnant grizzly is banging on people's windows, and then Severina, Bear's girlfriend, sees a group of moose running along the road. At the end of the story, Bear recognizes these bizarre animal activities as warnings that the animals are giving to humans about the coming of the zombies: "This is a mother slapping her hands against our windows with new life inside of her telling us that the murder of the world was coming. Those moose running away. *They were all trying to warn us*: things were going to get bloody and worse" (42, emphasis original). In the future of "On the Wings of This Prayer" that murder is in full swing: "The Hair Eaters have eaten all the caribou, moose, bear, fox, wolf, bison, buffalo and everything under the earth here. We are too scared to check nets as they feast on our catch underwater and wait for boats. All we have left are creatures of the air: ducks and ptarmigan, geese, swans. We eat roots and pretend they are what we used to love" (15). Of his choice of subject matter, Van Camp has said, "I write about what breaks my heart in the world and I 'fix' it in a story" (Rocan). While these two stories do not offer a "fix" for bitumen extraction, they do show characters striving to live in such a way that their *logos* accords with their *bios*, which might be read as an alternative to the lifestyles that lead to the arising of the Wheetago. Van Camp describes Bear as "my new gladiator and if anyone can defeat a Wheetago and walk a path that leads to forgiveness, it's him" (197). The story ends without making clear if or how he will walk this path, but it does offer the hope that it may be possible.

Ecologist Stan Rowe argues that humanity "makes a virtue of mining and releasing into air, water and soil the many poisonous materials that used to be safely sequestered underground: heavy metals, radioactive minerals, long-chain hydrocarbons such as coal and oil" (117). One response to this rhetorical management of human action is to work through the psychic consequences in story. In *Our Vampires, Ourselves*, Nina Auerbach writes, "what vampires are in any given generation is a part of what I am and what my times have become" (1). Her book focuses on the shifts in vampire representation alongside shifts in the political climate. More recently, Ted Atkinson has argued that the television series *True Blood* "extends the long-standing cultural practice of making vampires screens for projecting collective desires and anxieties" (213). Van Camp's zombies can be read through a similar lens. His stories show how people acting to fulfill their own desires become like zombies. The exploitation of bitumen is turning us into zombies, mindless consuming automatons—and the consumption of bitumen involves the consumption of other people, which is a result of "Wheetago culture," as some people begin to farm other people to serve the zombies and save themselves (Van Camp 15). As Stoekl puts it, "The self is a function of the world as standing reserve, the collection-disposal of accumulated raw material. And the self becomes raw material as well" (*Bataille's Peak* 139). Our appetites, desires, lusts, threaten to consume others; others' appetites, desires, lusts, threaten to consume us.

In "On the Wings of This Prayer," we learn that the first Wheetago was an old trapper who lived in "The Tar Sands of Alberta" but "no matter how much money the oil companies offered him, would not budge, so they built and dug around him" (10). This cuts him off from other people and he "quit coming to town. There was a family who went to visit him, to bring him supplies, but he had changed" (10). He is killed, dismembered, and burned before he can infect anyone else, but "the land was uncovered and turned for years by excavators, tractors and the curiosity of men. We think those machines must have moved the heavy rocks that covered his limbs. We think his fingers were able to crawl back to the torso and legs and head"

(10). This body, reassembled due to the extraction of bitumen, bites the mother of another family who was "in the way of the Tar Sands of Alberta" (9); although she asks her family, in her moments of lucidity, to kill her, they don't, and she escapes—"They think she was the first and they say she is still here as their queen, that she gives birth to them through her mouth" (10). In both of these origin stories, people who block bitumen extraction become isolated, infected, and insatiable: "In both stories it is the Tar Sands to blame. This is how the Wheetago will return" (12).

A review of the collection, posted online by L.A. Kranz, states that "the Wheetago are us, that monstrous part of our nature that rises with greed and unearths uranium, radium, plutonium, fossil fuels...This book is a Warning." It is more than a warning, though, because, in the characters of Bear, Severina, Snowbird, and Torchy, Van Camp offers models of heroic action in response to the Wheetago—a symbol of contemporary society's mindless consumption and waste—and this action does not approach the world as standing reserve.

The Comaroffs compare South African zombies with the strangeness of English Gothic fiction "where industrialization was similarly restructuring the nature of work and place" (794). They note, "zombie tales dramatize the strangeness of what had become real; in this instance, the problematic relation of work to the production of social being, secured in time and place" (794). They explain that the forces of global capital

> fracture the meaning of work and its received relation to place. Under such conditions, zombies become the stuff of "estranged recognition": recognition not merely of the commodification of labour, or its subjection to deadly competition, but of the invisible predations that seem to congeal beneath the banal surfaces of new forms of wealth. In their iconography of forced migration and wandering exile, of children abused and relatives violated, the living dead comment on the disruption of an economy in which productive energies were once visibly invested in the reproduction of a situated order of domestic and communal relations; an order through which the present was, literally, kept in place. And the future was secured. (795)

This could well describe the world depicted in "On the Wings of This Prayer," where a desperate remnant of the previous order throws a prayer into the past in a last attempt to redeem the future. Rather than succumbing to the selfishness of zombies, Bear approaches otherness as relation, even the otherness of the Wheetago, and thereby begins, at least, the act of redeeming the future that seems lost in the first story.

Bear is convinced, for example, even after his encounter with the Wheetago, that "Dean's still inside it" (36) and he instructs Torchy not to kill it. The story begins with Bear's assertion that he is "starving" (20) for Severina, which links him to the Wheetago, who "the more they eat the hungrier they become" (39), and suggests that his desire to save Dean is also a desire to save himself, and that the potential to save Dean may offer the potential to save humanity from our zombie ways. There is a hint that Snowbird may have been a Wheetago once: when Bear asks if Snowbird and Torchy can save Dean, Severina says, "They'll have to blind him" (39); later Severina tells us that Snowbird has not always been blind, which leads Bear to question, "*Was he a Wheetago once?*" (40, emphasis original). If so, there must be a way to recover. At the end of the story, Severina gives Bear a message from Snowbird: "If you save everyone and put yourself before the beast and walk away, then you will know how heroes are born" (41). Bear's heroism puts him outside the economy of consumption, at least momentarily. He clearly still has appetites and desires, but they have become secondary to the true freedom of necessity. As Žižek has it, "True freedom is not a freedom of choice made from a safe distance, like choosing between a strawberry cake and a chocolate cake; true freedom overlaps with necessity, one makes a truly free choice when one's choice puts at stake one's very existence—one does it because one simply 'cannot do otherwise'" (*Defense* 70). Bear stands and confronts the Wheetago in Dean because he chooses to do so *and* because he must.

Lust, Addiction, and the Possibility of Redress

C.D. Evans and L.M. Shyba's satirical novel *5000 Dead Ducks: A Novel about Lust and Revolution in the Oilsands* connects to Van Camp's stories via this focus on appetites, but the novel focuses on the perspective of those who lust after power, and get it, rather than those who suffer as a result of others' pursuit of their appetites or those who resist that pursuit. It is unclear if anyone in the novel achieves the freedom of necessity, although the underdeveloped character of Nimmo may be a candidate. The main characters in 5000 *Dead Ducks* are the most monstrous, though the monstrosity here is that of the vampire rather than the zombie. Instead of being mindless, devouring automata, the characters in the novel are cold, calculating solipsists. Evans and Shyba create an alternate universe to explore potential consequences of the deaths of five thousand ducks on a bitumen mine's tailing ponds.

With the election of Alison Redford as the leader of the PC Party and, therefore, the premier of Alberta in 2011, who claimed Peter Lougheed as one of her political mentors, the narrative of "responsible development" was reasserted in Alberta. Redford's ethos may have initially carried more weight with progressives than her predecessor Ed Stelmach's, whose statements on the environment were always read through the lens of his statement, early in his term as premier, about refusing to "touch the brake" on development. However, Redford also argued forcefully for the necessity of building pipelines to enable increased bitumen production and, theoretically, higher prices. Interestingly, in *5000 Dead Ducks: A Novel About Lust and Revolution in the Oilsands*, the failure of the provincial government—the members of which are referred to derisively as "farmers" (21)—to defend industry after the titular avian incident, and industry's inability to get the required federal permits to build pipelines to the West Coast and, thereby, pressure Montexian (i.e., US) consumers to pay more for bitumen, lead JB Mailcoat, CEO of Petrofubar Energy, to move to replace the government of Alberia. He declares, "Well, that's the last straw. Provincial politicians ain't doing fuck-all to protect our industry from the feds. Time to throw th' fuckers out. We're just not gonna take this shit, I'm tellin' you right now! In Candidia Francis [i.e., Quebec] they got one million separatists with ten cents each. In Alberia, we got seven separatists with a hundred million bucks each" (22). Perhaps this is reflective of why, given the rise of the Wildrose Party's rise to official opposition in Alberta in 2012, "responsible development," for the provincial government, continued to mean expanded development.

In the novel's universe, the country of Candidia is governed by a coalition, which—unlike the attempt by Stéphane Dion, Jack Layton, and Gilles Duceppe following the 2008 federal election to replace the minority Conservative government of Stephen Harper—has succeeded in taking power. Following this counterfactual creation of a federal government unsympathetic to the fossil fuels lobby, the challenge posed to industry by the duck deaths leads to the successful rise of an Alberian right-wing separatist movement, funded by industry, particularly Petrofubar CEO JB Mailcoat. At their first meeting, Mailcoat tells Fred Fiddler, soon to be leader of the Alberia Special Party, or ASP, "Me and my rich oilpatch buds, we'll be your money muscle 'cuz we just mine this joint like Jupiter's moon, make a pile, leave a mess, who gives a shit, and then, get the hell out and go back to God's country [i.e., the US]" (34). This movement—led by Mailcoat, lawyer Walt Semchuk, law professor Ursula Vere, Deputy Attorney General Baycell Sharpe, and fronted by megalomaniacal and sociopathic lawyer turned politician Fred Fiddler—also becomes involved in a plot to seize secret gold bought by former Alberia Premier Mad Ludwig with Heritage Trust Funds and stored in the "Arsehold bunker" (74). While this gold distracts from the novel's satirizing of unfettered bitumen exploitation, it serves the plot function of allowing the proto-fascists to buy their way out of trouble following the nuclear holocaust of northern Alberia. This tragedy occurs when dysfunctional mobile nuclear reactors, purchased from the FarEasters to power bitumen extraction projects, melt down.

The bodily drives of the individuals involved certainly colour the form the revolution takes, before it descends into nuclear disaster, anarchy, and chaos. While planning the separatist revolution that provides the main plot of the novel, we are told that "Imagining the revolution has made Walt lusty" (72); he states that, though there is so much work to do, "other appetites have also to be fed" (72). JB is perpetually swilling beer. Ursula, a stocking-fetishist from the start, becomes a drug addict. Most interesting to me, though, is the transformation of environmental scientist Dr. Mavis Wong into a member of ASP. Mavis discovers and reports the duck deaths and initially seems to hold strong environmentalist credentials.

While still struggling between her loyalty to ASP or her environmentalist past, Mavis visits the Petrofubar psychologist Dr. Travis Zapper. Dr. Zapper tells her, "Your anxiety is due to insecurity about being in a job that is far too much for you to handle" (115). Mavis replies,

> That's wrong. I paid my dues and then some down in Tapperlite. We cleaned up arsenic, mercury and all kinds of crap left over from turbo-capitalist copper mining and had successful results. But in the oilsands, fossil fuel environmental destruction makes Tapperlite look like a tropical beach party. Don't you see? Dead ducks are just the symptom. Think about the disease...I can't sleep. (115)

In this scene, the novel strives to move beyond the factual, material substance of bitumen to talk about what bitumen represents: bitumen as a symptom of a cultural disease, bitumen as money, bitumen as power to satisfy human appetites. When Mavis focuses on the fact of the dead ducks and the larger disease of which they are a symptom, she can't sleep.

Jennifer Sonntag has written of the 2010 Deep Water Horizon/BP disaster in the Gulf of Mexico,

> We must not let this latest event (we should not dignify it with the term "accident") turn into just another commercial for therapeutic soul-searching and plans to plan for a renewable energy future. We need to forgo the usual tepid reliance on belated commission "findings" of crimes committed. No criminal investigation will redress the loss of whole ways of life in the Gulf, nor the spilt blood. The redress will occur in our changed ways, or not at all.

The same could be said of the "sacrifice zone" in Alberta's boreal forest. On the other hand, the public is faced with continued industry and government factual relativizing of the issue. For instance, in 2008 bitumen accounted for 0.15 per cent of global greenhouse gas emissions (Alberta Energy, "Facts and Statistics"); bitumen is equivalent to some Californian and Venezuelan heavy crudes; and per-barrel emissions are decreasing (CAPP), etc.

Although Mavis rejects Dr. Zapper's claim that she is "overreacting" (115) and does not fill any of the prescriptions he writes for her, her pangs of conscience disappear when she visits JB Mailcoat, CEO of Petrofubar, and is given a promotion, Petrofubar stock, and a new Hummer. Money proves to be a powerful, and fast-acting, narcotic. She sets out on a shopping spree and sheds her cargo pants and Gore-Tex jacket, replacing them with a crepe dress, duck feather coat, and stiletto boots. She declares, "From now on, I'll either be dressed like this or in the uniform of the ASP Party" (124). The feathers of the coat suggest a successful capture of the ducks that provided them as a resource as opposed to the waste of those who died in the tailings pond (see Chapter 2 and Levant, "Shed No Tears"). While Mavis continues to struggle with her conscience, she never rejects the ASP revolution and its promise of satisfying all of her desires, despite its consequences for those from her past. The upward counterfactual of endless growth and prosperity overcomes her concerns for the people left out and the environment sacrificed.

She remains sentimental about her former colleague, now AntiTox activist, Nimmo, but acts to prevent his protest against ASP before the party takes power (151). Following the election of ASP to government and Alberia's secession from Candidia, Nimmo is imprisoned, "accused of high treason against ASP" (205). Mavis visits him in prison and claims, "I got caught up in this. I needed money—" (206) and, "JB and I work closely and I got sucked into this whole scene. I'm not proud of it" (207). Feeling guilty, Mavis, with JB's assistance, gets Nimmo out of jail and transports him to safety across a border checkpoint with Candidia. Following this episode she says to JB, "We can still take the moral high ground JB. We can't stop what's happening but we can try to carry on some of Nimmo's work" (210). However, this crisis of conscience is quickly overwhelmed by the nuclear disaster, and Mavis quickly shifts into pragmatic survival mode.

As an indirect satire, the novel creates "characters who make themselves and their opinions ridiculous or obnoxious by what they think, say, and do" (Abrams 277), and this ridiculousness exposes the irrationality of our current petroculture. *5000 Dead Ducks* is not so much, then, about bitumen as it is about what bitumen represents and the danger of following that symbolism to its logical conclusions. In contrast to industry and government claims, we need to stop focusing on just "the facts" and pay attention to the symbolism of bitumen. What does bitumen, and, indeed, petroleum in general, represent in our culture? Opposition to projects like Keystone XL and Northern Gateway suggest that the narrative of petroleum as freedom is starting to lose its rhetorical power. However, as Mavis Wong shows in the novel, it has certainly not lost all its power.

Further, the potential for power to corrupt is seen in the character of Ursula Vere: by the time of her grisly death she has not only shaved her entire body but has also added "tattoos of serpents and winged medusas on both shoulders—artwork that seeps up her neck and on up the back of her waxed dome" and "piercings in her nose, lips and eyebrows"(179). She wears "a bikini top and a ratty, torn body stocking with massive silver chains clamped onto grommets at the backs and fronts of her knees" (180). Like the Wheetago in Van Camp's stories, and addicts in general, the more these characters acquire, the hungrier they become. Perhaps, then, though it is sudden and unexpected, and, from a plot perspective, rather too convenient a way to eliminate an unruly character, there is a certain logic to Ursula's untimely and brutal end on the chainsaw of Flossy Kinderman, the jealous partner of Ursula's former lover Celeste Kinderman. Ursula, in her growing obsessions and addictions, becomes progressively more monstrous as the novel proceeds. Her murder, then, might be seen as justified. Though it is presented as a crime of passion on Flossy's part, it might also be read as the only way to stop an all-consuming monster. Unlike Bear's engagement with Dean in "The Fleshing," though, there is no attempt by anyone to save Ursula prior to her death. This difference indicates a fundamental distinction in the views of these texts. The response to the murder by Celeste and Walt is, in the moment, entirely pragmatic as they make plans to dispose of Ursula's remains. After Ursula's death Walt is haunted by her memory, but, as we will see, there is no suggestion that her humanity might be redeemed after death either. Although Van Camp's depiction of the post-bitumen world is far more detailed, and, therefore, horrifying, than Evans and Shyba's abstract sketching of nuclear apocalypse, Van Camp also manages to suggest the stance or comportment necessary to create an alternative future. Evans and Shyba have no heroes; as a result, their conclusion struggles to gesture beyond the self-interest that has determined the characters' actions throughout.

Following the bloody end of Ursula Vere, Walt Semchuk, whose real name is Vladimir Semchuk, reflects on his actions: "It doesn't say much for him that he couldn't keep his lusty passions in check to prevent the aimless path of this revolution that has much more now to do with greed and hitting the jackpot with the Arsehold Gold than with completing an exercise in curiosity about the human condition. Well, at least he can walk away, once and for all, from Vlad, his vampiric birthname" (191). However, this idea of freedom from his past is wishful thinking on his part; not only is he burdened by the memory of Ursula (and a red bowling ball bag containing her head), following the nuclear accident in northern Alberia, Walt and other high-ranking members of ASP flee to the Arsehold Bunker. There, Walt has plenty of time to reflect on his actions while plotting the means of escape from the crowd of hostiles that surrounds the bunker.

His reflections might be read in terms of Grant's statements about how reprehensible it is to write "false theory" (*Philosophy* 87). Walt reflects on his values, claiming, "He believes in the rational versus irrational; logic versus paradox; beauty versus ugly, and is wondering how far this bogus revolution would have gone had it not been for the evil Professor Vere and her hasty cocktail napkin of Big Lies" (191). Walt's list of values here, and its apparently stable hierarchy of binaries, has been undermined by the preceding action of the novel. Indeed, Walt's own actions suggest that he does not always privilege those things that his "rational" reflection tells him he should. He displaces his own responsibility for the revolution onto Ursula, and dismisses his own involvement as "an exercise in curiosity." However, the greed of the revolution that Walt now dismisses and despises was justified precisely by his "exercise in curiosity" and his hypothesizing about the human condition. His false theory enabled him to pursue his selfish desires. Indeed, the greed had been present from the beginning; as Ursula pointed out to Walt during their first planning tryst, "revolutions inevitably consume themselves so it's just as well we end up rich" (66). Had the characters simply plotted to steal the gold, and not linked that plot to a political philosophy, the consequences of their success or failure would have been much smaller. Further, the napkin of Big Lies is a joint document, starting with Walt saying, "Fred and the Inner Circle have to accept that the true path to power is in telling lies, big lies. Really Big Lies to be hammered home with incendiary speechifying at every opportunity, in every venue" (102). The "Number One" Really Big Lie is that "The feds are going to shut down the oilsands" (105). By propagating this lie, ASP is able to foment revolution and engineer Alberia's secession from Candidia because, like Albertans, they have been conditioned to believe that oil extraction will make everyone wealthy. Ursula states, "There are three conditions that will motivate Alberian citizens to be a mob [:]...a charismatic leader,...the current uncertain economic and political times...[and] there have to be identifiable groups that threaten the average Alberian... In Alberia the mob turning points will be political and economic, namely those Eastern bastards and the environuts" (68). While the language in *5000 Dead Ducks* is maintained at an extreme of *realpolitik* and manipulation throughout, it is not such a stretch from that used by both the federal and provincial governments to characterize environmental groups.[5] The difference is that, in the novel, this extreme rhetoric is not balanced by claims of economic and environmental sustainability: the implication is that claims of "responsibility" and "sustainability" are simply covers, and "behind the scenes" the discussions of those in power are much more transparently selfish. As in "The Fleshing," though, lusting after power has monstrous consequences.

The Morning After...

Atkinson connects the monstrosity of vampires in *True Blood* to British Petroleum's management of the Deep Water Horizon disaster in the Gulf of Mexico, suggesting some of the ways that both vampires and oil corporations disguise their monstrous natures:

> Russell [the vampire king of Mississippi in HBO's *True Blood*] cloaks his intense profit motive in the trappings of gentility, grandeur, and affected social graces. For this reason, in addition to others, the vampire king can be seen as both counterpart and foil to Tony Hayward, the CEO of BP whose hapless attempts to maintain a persona of graciousness,

> affability, and compassion against the backdrop of environmental catastrophe made for a kind of absurdist theater of gentlemanly deportment in the mediascape. The comparison gains strength from taking into account Karl Marx's often-cited use of vampirism as metaphor: "Capital is dead labour that, vampire-like, only lives by sucking living labour, and lives the more, the more labour it sucks." (225)

It is worth mentioning at this point that labour is noticeably absent from *5000 Dead Ducks*. The only scene that recognizes the labour involved in bitumen extraction at all occurs at the start of the fourth and last section as the nuclear reactors begin to malfunction: "George Bovich is more interested in his bologna sandwich than in the workings of his portable mini-power reactor" (219). The action of the novel is located primarily amongst those who control the dead labour of capital, or want to. Atkinson continues:

> Žižek's adaptation of Marx yields further understanding of how Hayward's media persona renders him every bit as undead as his imperial counterpart in *True Blood*. Žižek explains how the effort to transcend clichés permeating public discourse can yield an unintended outcome in subjectivity: "not a person capable of expressing themselves in a relaxed, unconstrained way, but an automated bundle of (new) clichés behind which we no longer sense the presence of a 'real person.'" The figure that emerges is "a true monster, a kind of 'living dead'" embodied in the vampire. The vampire capitalist envisioned by Žižek finds almost perfect expression in the monstrous, undead gaffe that became Hayward's signature line: "I'd like my life back." (225)

As an attempt to garner sympathy for his own hardships in having to take responsibility for the disaster, Hayward's statement is clearly thoughtless in ignoring the far more terrible hardships caused by his company's failings for the families of the workers killed, for the livelihood of those who labour in the Gulf, and for all of the non-human others killed or sickened by the spill. However, it is worth asking what the ethical response should be to the suffering, not just of the innocent, but also of the guilty. *5000*

Dead Ducks raises this question, as the principal characters are, with the possible exception of Mavis, almost entirely unsympathetic. In the end, they are threatened by a faceless mob and forced to flee for their lives. Despite their responsibility for the disaster, though, readers may end up feeling more sympathy for them than for the faceless others outside the Arsehold Bunker, whose suffering is surely worse but is not described.

At the end of *5000 Dead Ducks*, as the three survivors prepare to flee across the Candidia/Montexia border with the assistance of the War Bonnet First Nation, to whose reserve they have managed to escape, Mavis says, "We have much to atone for, JB, but let's do our best to make it good. I don't know about Walt" (261), to which JB replies, "He's gonna have to live with himself" (261). It seems that these characters have awoken from the intoxication of money and power that directed their actions earlier in the novel. Now, perhaps like Tony Hayward, whose line about wanting his life back might also be read as a recognition that he cannot recover the illusion in which he previously lived—the illusion that the actions of his company were beneficial to those involved with them—these characters will need to find ways to live with themselves and the actions they perpetrated while intoxicated. Similarly, in "The Fleshing," if Dean can be redeemed from his Wheetago existence, he will need to learn to live with himself and the crimes he has committed. Ultimately, the final lines of "On the Wings of This Prayer" speak to the readers of these stories as they suggest the complicity of our lives, of our culture, with those of the characters who are lusting after power, greedy for money, intoxicated with dreams of independence: we read, "You *can* change the future" (17, emphasis original). "Now wake up" (17). Perhaps Hayward, in expressing his wish for his life back, had awoken, from the dream of endless growth and prosperity, to the nightmare that pursuing that dream has created and was expressing longing for a return to his past ignorance. As Stoekl notes, such a reality is "too horrible to think about" (*Bataille's Peak* xi). Like many of us, upon waking from a dream, or, perhaps, waking with a hangover and the memory of the intoxicating night before, Hayward's poorly worded wish for his "life back" can be read as a version of "I'd like to go back to sleep." Nevertheless, going back to sleep, ignoring the reality of depleting energy sources is a temptation, an upward counterfactual, that must be resisted, or, better, countered with a different temptation.

The temptation to return to the comforting dream of limitless growth, to the upward counterfactual of endless progress, is opposed in Van Camp's stories by the representation of the downward counterfactual of the future consequences of such continued sleep in "On the Wings of This Prayer." This temptation is also opposed by the upward counterfactual of the heroic actions of Bear, Severina, Torchy, and Snowbird in "The Fleshing"; there is hope that these characters, in adopting the right stance towards the threats they face, can deal with these threats and protect or save the future.

The conclusion to *5000 Dead Ducks* is less satisfying to me. While preparing to leave the War Bonnet reserve, JB trades one of his wooden duck decoys for a boy's dog, saying, "If ya put 'im in the pond, he'll float" (262). The image of a duck floating on a pond may offer a hopeful contrast to the sinking of the bitumen-coated ducks from earlier in the novel. However, this is a wooden decoy, not a real duck. It is being traded by a white man to a Native boy for the boy's dog, and the exchange is taking place as the architects of Alberia's destruction escape to freedom across the border while the Natives remain to live with the fallout of that destruction. Rather than suggesting the beginnings of atonement, it suggests a continued selfishness, as JB finds a new source of comfort for himself even while fleeing justice.

As I noted in the Introduction, Rob Nixon asks, "How can we convert into image and narrative the disasters that are slow moving and long in the making? How can we turn the long emergencies of slow violence into stories dramatic enough to rouse public sentiment and warrant political intervention?" If Van Camp's stories manage to address the challenge Nixon poses by representing the slow violence of today as fast violence in the future (today's bitumen extraction will be tomorrow's zombie apocalypse), a shift that requires action today on the part of the readers (we must "wake up"), Evans and Shyba's novel fails to effect a similar conversion. *5000 Dead Ducks* does turn the slow violence of bitumen extraction into the fast violence of nuclear meltdown, but the novel ends with the suggestion that the principals can still act in the future to make amends for what they have done. By contrast, in “On the Wings of This Prayer” there is nowhere else to go; there is no place free from the consequences of ecocide. “There is no planet B,” as the slogan has it; the future is lost, and it is only in the past that humans are free to change that future.

Conclusion

Living (with) Bitumen

UNSUSTAINABLE OIL BEGINS, in Chapter 1, and ends, in Chapter 6, with things buried in Alberta's bituminous sands: an angel and a Wheetago. One, whole, proclaims the glory of the natural world; one, dismembered, declares the horrors of despoiling that world in pursuit of human appetites. Both call on readers, through the insertion of creatures in the sand, to question the status quo. Both call on readers to move from a passive acceptance of the current system of bitumen extraction to an active questioning of that system. Indeed, that is the role *Unsustainable Oil* shows fictional literature about the bituminous sands playing in general.

My second time on the Suncor tour, in September 2012, another version of something buried in the sands was presented: the film *The Suncor Ankylosaur*. This time the creature is factual. The film tells the story of a fossilized ankylosaur discovered in a Suncor mine and how the company and its workers deployed their resources, equipment, and ingenuity to "rescue" the fossil from the sands. Unlike the superintendent in Wiebe's story who worries that what Tak has found in the mine will require shutting down production and bringing in paleontologists, *The Suncor Ankylosaur* shows a corporation and its employees who are co-operative and accommodating, going above and beyond to help the scientists from the Royal Tyrell Museum who arrive to excavate the dinosaur. At the same time, Project Manager Doug Lacey, who led Suncor's excavation team, also notes in the film that "we were able to sustain our day-to-day production and operation and have a good event happen right beside it and not impact it." The industry depicted in the film is one that has grown beyond the tenuous early days described in Wiebe's story. It has grown to the point where uncovering a fossil becomes a public relations opportunity for Canada's largest energy company (Lewis), with a film crew on hand to document the fossil extraction as it happens. The fossil extraction occurs simultaneously with the daily bitumen extraction, that banal extraction of other decomposed organisms lacking the charisma of a dinosaur. The excitement shown by the Suncor employees working on the ankylosaur extraction suggests a lack of similar relation to the organisms whose remains comprise the bitumen.

The ankylosaur is described as one of the most important fossil discoveries in Alberta's history, so perhaps it is appropriate that it has been named—the Suncor ankylosaur—after the first large-scale commercial operation in what has become Alberta's most important industry. It might have been named differently: the Clearwater ankylosaur, for example, after the geological formation in which it was discovered. The film shows the fossil becoming a symbol, like the bison discussed in Chapter 2, for the industry: it is *because of* the industry that this fossil was "saved."

The blog *McMurray Musings*[1] reiterates the point of the film: "Were it not for the oilsands operations this fossil would never have been uncovered, and would have remained in the earth for all eternity. Instead it has been found and will add immeasurably to our understanding of ankylosaurs, and of our general understanding of creatures from that time so incredibly long ago." Unlike Wiebe's angel or Van Camp's Wheetago, whose discovery in the sands questions the mining of bitumen, the film on the Suncor ankylosaur justifies that mining. *McMurray Musings* continues,

> There is yet another reason this find is so incredible, though—were it not for where it was found it would never have been excavated. The fossil sat high in the wall, beyond easy reach, and often when fossils like this are found in other places they are left there because it is simply too difficult to extract them. The resources required to excavate them (large machines and skilled operators) are too far away and too costly—but in this case good fortune prevailed. Suncor, of course, has all the large machines and the most skilled people to operate them, and what could have been a major find that would never be seen by the world instead became one that could be, and was, recovered.

It is "good fortune" that Suncor exists so that this precious fossil could be found and can now add to human understanding. The film *The Suncor Ankylosaur* is another version of the narrative of benefits of the petroleum age: massive investments of energy lead to and enable human progress by paying for education, art, culture, and science. The film does not quantify the energy needed to extract this fossil, but it does make the point, as does the *McMurray Musings* blogger, that under other circumstances its extraction would have been too costly. Certainly, in a world without the cheap energy of petroleum such an undertaking would have been too costly. My point though is not that its extraction was "wasteful" and therefore bad; rather, the representation of its extraction is used to justify another form of extraction that views the world as object, as standing reserve. The ankylosaur becomes a symbol for bitumen extraction and all the wonders that bitumen extraction makes possible. Without bitumen, as standing reserve of energy, the ankylosaur would have remained "lost"—as though its existence is only meaningful once recognized by humans.

Coming back to the question of sustainability and its impossibility for non-renewable resources that I raise in the Introduction, narratives like that of *The Suncor Ankylosaur* begin to look different. When the facts—this fossil was discovered and extracted in the Suncor Millennium Mine by scientists from the Royal Tyrell Museum with the assistance of Suncor employees—are set beside counterfacts—bitumen mining will uncover an angel, or cause the death of innocents, or result in the eternal suffering of Elder Brother, or awaken a Wheetago, or foment revolution leading to nuclear meltdown—the facts begin to look different. Rather than pride in human accomplishment, the facts may suggest hubris: the potential to extract the fossilized remains of dinosaurs is a temporary cultural phenomenon resulting from the prosperity of this historical moment.

McMurray Musings, by contrast, builds on the facts with a different set of counterfacts:

> I had absolutely nothing to do with it's [*sic*] discovery or recovery, people, and yet I am so proud of Suncor and their employees. Suncor saved this little ankylosaur from being lost in rock forever, unknown, undiscovered, and unappreciated. I can imagine some day when it goes on display and it is surrounded by small children who gaze in wonder and excitement at it's [*sic*] size and strangeness. That image alone makes me smile, people, and I bet it makes you smile, too. This is one momentous discovery we can all be proud of, Fort Mac. This little ankylosaur belongs to all of us—and we have Suncor and their employees to thank for bringing him home to us.

This passage strikes so many of the notes from the justifications for bitumen extraction—the redemption of the past (saving the ankylosaur from being "lost forever"), the imagined ideal future of "wonder" for small children, the benevolent smiles of those who secured this future for those children through their work, the invocation of "home"—but these upward counterfactuals ignore any possibility of downward counterfactuals. They also ignore the counterfacts of the ankylosaur's recovery: getting to the stone that encases the fossil, as indeed all bitumen mining does, required removal of several metres of overburden. "Saving" this prehistoric specimen "for future generations" requires the sacrifice of the habitat of present creatures, contributes to the region's "ecocide." As discussed in Chapter 2, acting in the service of an other—in this case the ankylosaur—requires the sacrifice of other others.

At the end of *In Defense of Lost Causes*, Žižek offers a prescription for preventing ecological destruction through imagining the worst possible outcome as guaranteed:

> For Badiou, the time of fidelity to an event is the *futur antérieur:* overtaking oneself towards the future, one acts now as if the future one wants to bring about is already here. The same circular strategy of the *futur antérieur* is also the only truly effective one in the face of calamity (say, of an ecological disaster): instead of saying "the future is still open, we still have time to act and prevent the worst," one should accept the catastrophe as inevitable, and then act to retroactively undo what is already "written in the stars" as our destiny. (*Defense* 460)

This may be read as the logic of speculative fiction in general—as expressed in Ray Bradbury's line, "The function of science fiction is not only to predict the future...but to prevent it" (Sturgeon viii)—and can be seen in the hope embodied in the counterfactual insertions in the literary texts discussed in *Unsustainable Oil,* operating most clearly in "On the Wings of This Prayer." While Žižek points to examples of how such pessimism has worked to bring about change in the past through counterfactual imaginings of the future—in Horkheimer and Adorno's *Dialectic of Enlightenment* or those who predicted the Cold War would be lost by the West—and I am generally persuaded by his assessment of our circumstances, I am not sure who this "one" is who acts, and what this "one's" motivation is in circumstances that are already lost. Perhaps this "one" is the reader who, reading in isolation, is opened to the recognition of true freedom as necessity and the change in stance towards the world that requires (see Coleman). Even in circumstances that are lost, as George Grant has it, it "always matters what each of us does" (qtd. in S. Grant 98). Further, if what each of us does is to recognize our interconnectedness and change our actions accordingly, it may be possible to retroactively undo our collective "fate."

In "With an Eye toward the Future: The Impact of Counterfactual Thinking on Affect, Attitudes, and Behavior," Faith Gleicher and her co-authors summarize psychologists Shelley Taylor and Lien Pham's distinction between outcome simulation, "in which the individual imagines a (usually desirable) outcome," and process simulation, "in which the individual imagines the pathway by which he or she can arrive at that outcome" (300). In Taylor and Pham's experiment, students imagined getting an *A* on an exam, the process of studying for the exam, both, or neither. Gleicher and her colleagues explain,

> although participants who simulated the desired outcome of receiving an *A* reported greater motivation to study for and do well on the test, the students who simulated the *process* actually were more confident of their ability to succeed on the test, studied longer, and performed better. According to Taylor and Pham, the process simulation led to these effects because it helped participants develop a plan for studying that they could then follow. (Gleicher et al. 300)

Here the difference between positive and negative counterfactuals returns. As discussed in the Introduction, Gleicher and her co-authors note that "upward and additive counterfactuals are more likely than downward or subtractive counterfactuals to lead to plans for future behavior" (301). They continue, "Because of the power of counterfactual simulations to stimulate emotional responses and because of their inherent process component, they may be particularly effective in spurring individuals to engage in further simulation that directly influences behavior" (302). It seems to me, however, that the upward counterfactual narratives are on the side of the status quo and are more likely to influence behaviour. We want to imagine the future as one full of wonder for our children, and, therefore, to save the ankylosaur. Arguing from "the facts" that continued petroleum extraction will lead to economic growth and prosperity for all offers a more appealing counterfactual than arguing from "the facts" that the future is already lost and that we must impose worldwide rationing of energy expenditure, "ruthless punishment of all who violate imposed protective measures," "large scale collective decisions," and "the reactivation of...the 'informer' who denounces culprits to the authorities" (Žižek, *Defense* 461). Missing from such a totalitarian formulation is the possibility of joy.

Stoekl's reading of Bataille offers a more hopeful vision of the future even if it, too, accepts the inevitability of energy decline. This decline, though, is unthinkable:

> What is clear is that one kind of matter, one energy, one plenitude, is dying; another, monstrous, already here, already burning, announces itself...Up until now, the development of thought, of philosophy, has been inseparable from the fossil-fuel powered growth curve, from "civilization." The downside of the bell curve is non-knowledge because the event of the decline of knowledge, the disengagement of philosophy from economic and social growth, cannot be thought from within the space of knowledge growth (the perfection of modern truth) or its concomitant absence. We are in unknowable, unthinkable territory—an era of disproportion, as Pascal might call it. The era of Bataille's peak. (204)

This territory—rather than operating in the *futur antérieur,* in the counterfactual worst of all possible worlds—strives to acknowledge our inability to know even the facts. Rather than acting based on "the best available science," we must act despite that science; that is, even though we know that science is incomplete, even though we know that we do not know, we must still make decisions. However, when those decisions begin from a recognition of our non-knowledge, our stance towards the world should be different, and, thus, so should our "necessary calculating" (Grant, *Technology and Empire* 35). Stoekl continues, "For Bataille, any assurances concerning the future, either good or bad, were beside the point, even silly; instead, there was the play of chance, the affirmation of what has happened, what will happen. The left hand spends, in gay blindness as well as science, and the future is affirmed, in the night of non-knowledge" (204). This affirmation of what will happen is not the same as Žižek's acting as though a catastrophic future is already present. Such an extreme downward counterfactuality seems certain to lead to either despair or the continuation of our culture of denial if it is not linked to some form of hope: Bataille's joy, blind as it is, offers some hope. In "The Fleshing," we see hope in the heroism of Bear and the community of resistance forming against the Wheetago; we can also see it in the promise of an intimation of perfection offered to Tak in "The Angel of the Tar Sands": "Next time you'll recognize it...And then it'll talk Japanese" (191). For Stoekl, the non-knowledge of energy decline should not, he says, lead to despair or inaction because

> we know the difference between sustainability and catastrophic destruction; we know the difference between global warming and a chance for some, even limited, species survival. But we also recognize, with Bataille, the inseparability of knowledge and non-knowledge, the tilt point at which, rather than cowering in fear, we throw ourselves gaily into the future, accepting whatever happens, embracing everything, laughing at and with death. We will a return of recalcitrant bodily and celestial energy, or the sacrifice of the logic of the standing reserve; we bet against the vain effort to will and endless autonomist freedom. (204)

Such a bet will affect our actions in the world, but, more importantly, it will change our stance in the world, our comportment towards the world, and that shift in relation will not lead us to Žižek's prescription for change. Ultimately, for Stoekl, "after Bataille, we refuse to take the downside of the bell curve as a simple and inevitable decline into feudalism, fundamentalism, extinction [and perhaps we can add, contra Žižek, totalitarianism]. We understand all that depletion implies, and we embrace it" (205). Stoekl explains in a footnote that

> what society will end up having to affirm...is not simple waste—fossil fueled production and destruction—but the burn-off of energy in the movement of an intimate world (muscles in agony and ecstasy, for example). If society's "freedom" is reoriented toward such an expenditure, then the other freedom, that of simply consuming (*consommer*), can be left behind. If such a reorientation can be affirmed, then conventional recycling can be something more, or other, than a mere sign of "personal virtue." (234)

It seems to me that this version of what "we understand" has more potential as a counterfactual than Žižek's. The literature explored in *Unsustainable Oil*—despite the prevalence of downward counterfactuals, despite the likelihood of things in the future being in many ways worse than they are now, of the continued exploitation of the bituminous sands leading to social and environmental problems—searches for, tries to find, and points towards alternative affirmations: affirmations of intimate relationships between those—humans and non-humans—who survive, affirmations of the sacrifices made by those who sought to find alternatives to the logic of liberal capitalism. These texts argue for the reorientation Stoekl describes. After the exponential growth culture enabled by fossil fuels, on the downside of the bell curve, alternatives to the objectifying rational-technocratic view of the world again become possible.

In contrast to the "just the facts" narrative of industry and government, then, it is important to consider and compare the counterfactual narratives generated by all sides in the bitumen debate. Counterfactuals are a necessary counterbalance to the dominant discursive counterfictionality that operates in discussions about bitumen, because such narratives always operate in the subtext of "factual" arguments, and foregrounding the counterfactual can raise questions about what is normally taken for granted or marginalized by dominant discourse. The challenge, though, that we're left with is not countering government and industry discourse with better facts or better interpretations of the facts but in generating upward, social counterfactuals that we can embrace and affirm, given the cultural-ecological crisis that we inhabit. We need to find ways to laugh at and with death in these good, and always already sufficient, lands (Kaye). Reading fictional literature can help with this reorientation.

Epilogue

Whitney Lakes Provincial Park

Looking from the cabin toward the lake: there is a row of pine trees on the left (whose dead lower branches I gather for kindling), a beautiful forking paper birch in the middle (its white bark peeling in whorls to show flecks of brown and black beneath), and the boat house and some scraggly poplars on the right; the grass descends into a beach mixed with sand grass (which we've been pulling out for years), then the pier, the boatlift, the boat, the water; then more water and the farther shore.

September 10, 2013: The Alberta government announces a new wetland policy that moves away from the "no-net loss" principle, will rank the relative value of wetlands, and will allow companies to pay into an education fund rather than create new wetlands to compensate for damaging or destroying others.

I remember picking saskatoons, walking into a hornet's nest and being stung eleven times on my right arm...

As a Canada Centennial project, my paternal grandfather, grandmother, and their children built a cabin on Laurier Lake. In 1982, the neighbouring lakes—Whitney, Borden, Ross—and the northwesterly portion of Laurier Lake were made into a provincial park: Whitney Lakes Provincial Park. I have swam, fished, water-skied, canoed in Laurier Lake nearly every summer of my life. I have hiked the trails and picked saskatoons, blueberries, dewberries, and wild strawberries. I wrote chunks of my dissertation there. I wrote chunks of this book there. My sons are the fourth generation of my family to spend time there. When I think of places that matter to me, I think of my family's home in Edmonton, I think of my parents' farm, and I think of the lake.

I remember going to get water from the red pump in the trees, then, when it dried up, to the blue pump up the road. I remember Dad carrying the pail, holding my hand with his free one...

In Warren Cariou and Neil McArthur's film *Land of Oil and Water*, Cariou visits a site of personal and familial significance near Meadow Lake, Saskatchewan, a place that could be threatened by expanding bitumen operations. In the summer of 2012, while at my family's cabin on Laurier Lake, one afternoon we went for a drive in an effort to get our two-year-old son to have a nap; my wife and I saw a sign for a Lindbergh oil sands operation just minutes north of Whitney Lakes Provincial Park. Even though I had been studying this development for several years, had seen the maps showing where the deposits are, I was surprised to see an active operation that far south. A look at Alberta Energy's Oil Sands Projects Map reveals that there are several operating bitumen projects adjacent to Whitney Lakes.

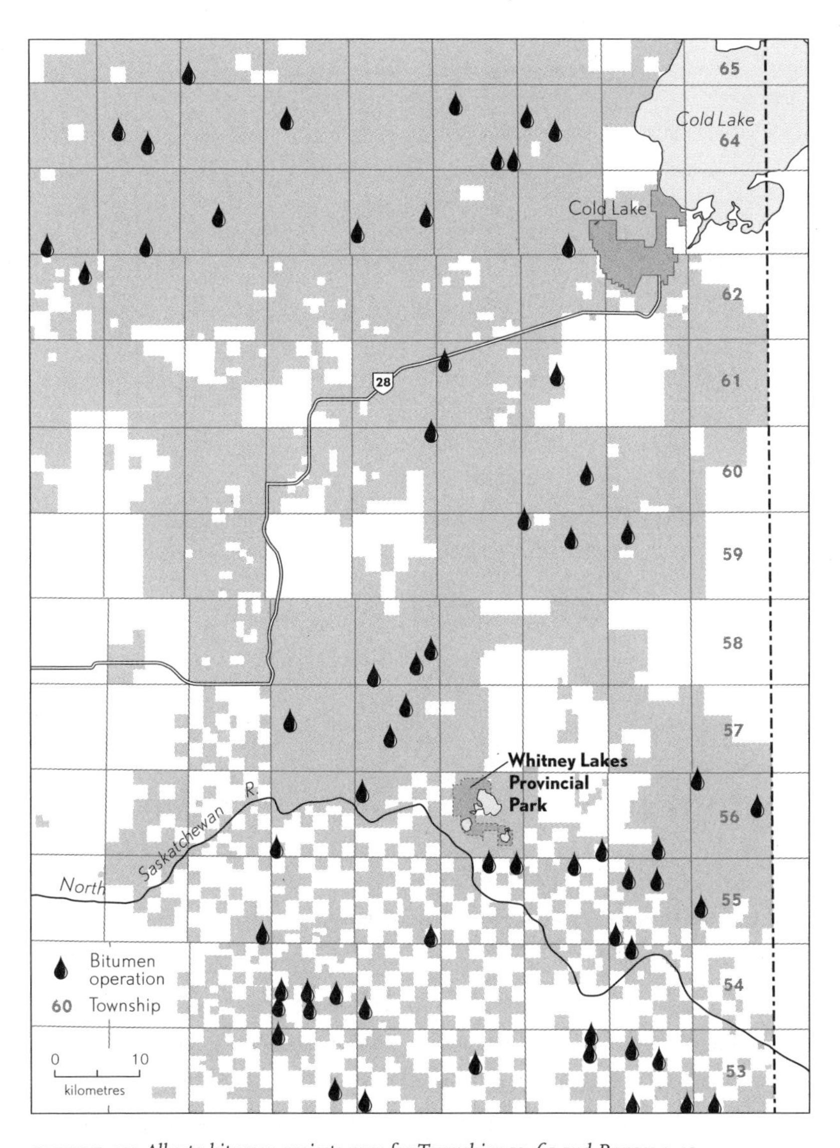

FIGURE 13: Alberta bitumen projects map for Townships 53–65 and Ranges 1–10.

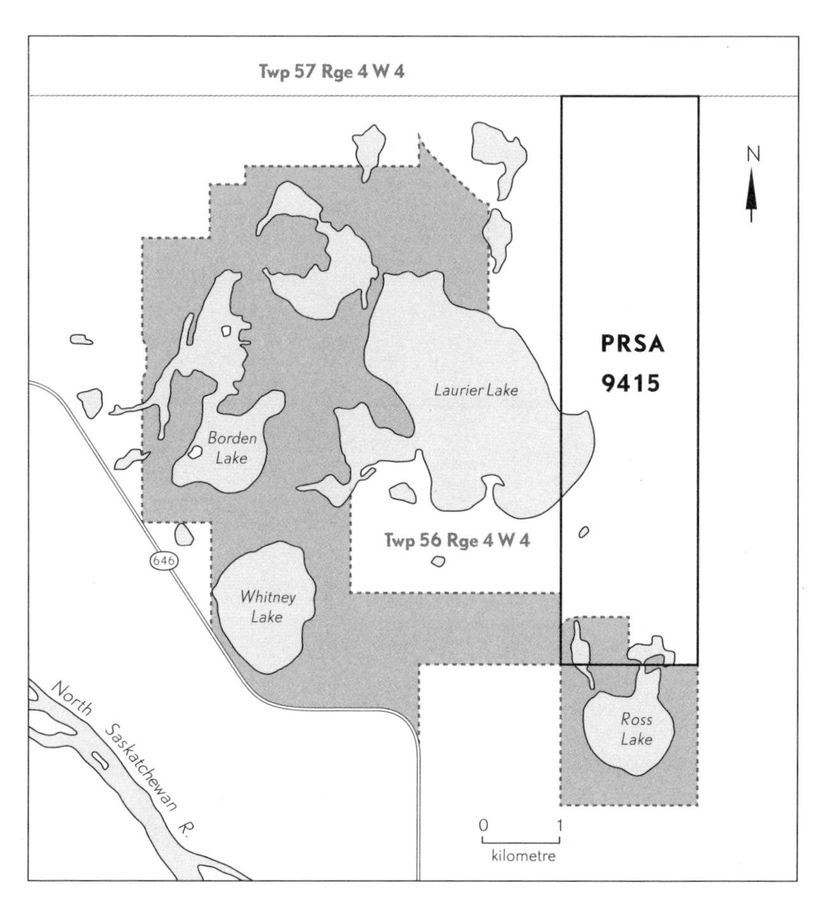

FIGURE 14: Proposed PRSA 9415 in relation to Whitney Lakes Provincial Park (shaded).

Spring 2013: several emails were forwarded to me regarding Canadian Natural Resources Limited's proposed Primary Recovery Scheme Amendment (PRSA) 9415, which would operate on the sections immediately east of Laurier Lake and north of Ross Lake. The proposal includes the east side of Laurier Lake and the north end of Ross Lake, raising the potential for horizontal wells to extend under the lakes.

I remember the water receding every year until we couldn't put the boat in anymore...

Now the water has started to come back and CNRL has proposed a Primary Recovery Scheme to get at bitumen beneath the lake.

I began the work that has led to *Unsustainable Oil* amid the environmentalist rhetoric that declared, "if this resource was located an hour north of Edmonton or Calgary it wouldn't be happening." I imagined writing something that would help people to see how we are connected to what happens in other places, and in Northern Alberta specifically. I did not imagine development of this resource would threaten a place to which I have a deep personal connection. I argued on the basis that people's connections to places matter and that the decisions regarding what should happen in those places should be made by the people who care deeply for those places. I still think this. The government's term—"stakeholder"—does not capture the primary, personal, material, embodied stake that people have in the places where development is occurring. This is, at least part of, what Angus means by "inhabitation."

James S. Edwards reminisces about his experience with Karl Clark outside of his lab:

> During the 50s I had the privilege of knowing him not as a scientist but as a friend and good neighbour of my parents. Both our families had summer cottages at North Cooking Lake, and the Clarks' cottage was offshore on Waverley Island. They parked their car on our property and launched their boat from a little jetty that Dr. Clark had built on our shoreline. I did not see him as the historical figure he later became but as a cottager who, like my parents, was devoted to simple country life for which chopping wood and attending to the many chores demanded, was a respite from city life. I remember being assigned the chore of ferrying a boatload of church ladies, friends of Mrs. Clark's, to Waverley Island. On arrival, we were greeted by Dr. Clark, scythe-a-swinging, clearing a pathway through the reeds to his family dock. (x)

Clark's correspondence reveals his love of this place, and one wonders about the path his career would have taken had the bituminous sands been located on the shores of, or underneath, North Cooking Lake.

George Grant writes of Darwin:

> What is the relation between his "loving recognition" of the beasts and flowers and what he cognised about them? It may well be said that what he knew about them means that no educated person should ever recognise them again in the same way people once did. But the question remains: is Darwin's love of the beings expressed in what he discovered about them? This is not to deny that he must have rightly loved his own remarkable act of synthesis, but that is a different thing from his love of the beings themselves. What was the relation of his love for them to what he discovered about them (and indeed about us)? To love something with intelligence is to want it to be. What is it about the animals as the product of modification through natural selection which would make us want them to be? (*Technology and Justice* 64)

We might ask, similarly, of Clark—who was, no doubt, rightly proud of his remarkable act of analysis that succeeded in separating bitumen from the matrix in which bio-geo-chemistry (call it if you will, nature) has locked it, and who clearly loved the place in which bitumen was deposited—is there anything about his discovery that relates to his love of the place in which it was discovered? Is there anything about his discovery that relates to his love of other places, such as Waverley Island? It seems to me that for Clark, what we see in his life is representative of the culture in which he lived (and of our own): a separation of the intelligence from love. This separation may be necessary for survival in this culture—it is too painful for us to want the things that our actions destroy to continue to be; it is too painful to combine intelligence with love. Instead, we separate the action of our intelligence from the things and the places we love, we talk of work/life balance, separating our existence into these parts; we move our objective analyses into laboratories in cities and ignore the consequences of that analysis for the wilderness where the objects belong, where they are part of an ecosystem and part of a history of inhabitation.

How can I describe this place to suggest what it means to me? The quality of the light? The way the world darkens when a storm rolls in? I don't know that I can say anything original about it. "Whereof one cannot speak..." but I feel compelled to try.

Then there are the gas wells behind, the lease roads where the ATVs *go now that the water has come back up—"At least they're off the beach," Mom would say...*

"Locals" are not the only ones who have a stake. I don't live in Fort McMurray. I don't live at Laurier Lake. Many of those opposing PRSA 9415 have primary residences elsewhere. It would be easy to dismiss this opposition as a well-off elite who doesn't want to see the property values of their vacation homes decline. But that would oversimplify things, would exclude those who live there full-time and are by no means well-off.

I recognize the privilege that enables me to drive three hours east of Edmonton to stay in a cabin and write, hike, and swim. I recognize the inadequacy of arguing in defense of this privilege against those whose livelihoods depend on diminishing the sentimental, aesthetic, or even ecological value of that place. Nevertheless, I am compelled to make such an argument, regardless of how inadequate. Latour states, "Our quandary is similar to that of a non-violent pacifist who still wishes to be 'stronger' than a violent militarist. We are looking for weaker, rather than stronger, explanations, but we still would like these weak accounts to defeat the strong ones" ("Explanation" 165). In the current hegemony, it seems impossible to argue, on "pragmatic" grounds, that one place or one constituency should overrule the good of the majority. Such arguments are dismissed as

NIMBY-ism. However, it should also be possible to argue that ecological interconnections are threatened by damage to individual places and that ecological integrity should trump any individual energy extraction project.

The response to past dispossessions that have created inequalities is not continuing dispossession; the response is not destruction of remaining wild places (disturbed and stressed as they are) in the name of economic progress and the lie of equality that such progress is supposed to bring. It is people living in places that matter to them, inhabiting those places, despite the impurity of that inhabitation, despite the inextricable linkages between that inhabitation and various forms of dispossession. Some of the people opposing PRSA 9415 work for CNRL and clearly recognize the complexity of their position.

CNRL has, at the time of this writing, for the moment anyway, withdrawn the application for PRSA 9415. I know of no relationship between local opposition and this withdrawal and mean to imply none. As with Dodds-Round Hill (see Chapter 1), CNRL, or another company, may well file a new application at some point in the future.

I remember helping Dad fill the boat up with gas: holding the five-gallon pail, one foot on the boat, trying to balance as the fluid sloshed about and the boat rocked in the water; the smell of the fumes, the rainbow sheen on the water as the spillage floated away...

Notes

Introduction

1. Early on, the working title of this project was *Untoward Sustainability*, which inverted the title of the report of the 2008 House Standing Committee on Natural Resources: "The Oil Sands: Toward Sustainable Development." Later, it was *Untoward Bitumen*. Both titles played on the following meanings of *untoward*:

    1. a. Not having or showing inclination, disposition, or readiness to or for something; disinclined.
       b. Showing lack of proficiency or aptitude; inept, slow. *Obs. rare.*
    2. a. Of persons (or animals), their disposition, etc.: Difficult to manage, restrain, or control; intractable, unruly, perverse.
       b. Of things: Difficult to manipulate, work, deal with, or perform; stubborn, stiff.
       c. Awkward, clumsy; ungainly, ungraceful.
    3. Characterized or attended by misfortune, calamity, vexation, or annoyance; unlucky, unfortunate, ill-starred.
    4. Unfavourable or adverse to progress; unpropitious, unprosperous.
    5. a. At variance with good conduct or propriety; indecorous, unseemly, improper; foolish.
       b. Marked by lack of reason or fitness. *Obs.*

2. In 2013, Suncor launched their "What Yes Can Do" campaign with television commercials, YouTube videos, and an extensive website. It makes essentially the same point: anything is possible if we just say "yes" to development. See www.whatyescando.suncor.com.
3. George Grant has argued that "All coherent languages beyond those which serve the drive to unlimited freedom through technique have been broken up in the coming to be of what we are. Therefore it is impossible to articulate publicly any suggestion of loss, and perhaps even more frightening, almost impossible to articulate it to ourselves. We have been left with no words which cleave together and summon out of uncertainty the good of which we may sense the dispossession" (*Empire* 139). Given that *Unsustainable Oil* is an attempt to argue that literature can still intimate, at least, such a sense of dispossession, I disagree with Grant's certainty on this point. However, I agree with him that it is extremely difficult; perhaps, as I claim in the Conclusion, following Allan Stoekl's reading of Georges Bataille, as we stare into the face of energy decline, and the unthinkable sufferings that decline will bring, perhaps the articulation of alternatives to the unlimited freedom imagined by techno-capitalism again becomes possible.
4. In a pamphlet called "Upstream Dialogue: The Facts on: Oil Sands," the Canadian Association of Petroleum Producers includes a graph from the International Energy Agency (IEA), which shows global energy use increasing during the period 1990 to 2035 from approximately nine billion tonnes of oil equivalent to approximately sixteen billion tonnes. Superimposed on the graph is the statement, "Oil sands help supply global energy needs" (11). This graph implies that the IEA prediction for global energy demand is a "fact" and that oil derived from bitumen is required to meet this demand. The logic for bitumen extraction, indeed for liberal modernity, despite its foundation in a belief in human freedom, depends on interpreting a certain set of facts about the present as necessarily leading to a definite outcome in the future. *Unsustainable Oil* argues that interrupting and questioning this narrative can re-open the indeterminacy of the future.
5. Das and Chapman were co-directors of Cambridge Strategies Inc., an Edmonton-based consulting firm, before Chapman moved on to become executive director of the Oil Sands Developers Group, and then, in 2013, executive director of Northern Initiatives for the Edmonton Economic Development Corporation, for whom he blogs as "Oil Sands Ken." His first post states his passion for "finding new ways of thinking about the oil sands." Das continues as principal of Cambridge Strategies and is author of *Green Oil: Clean Energy for the 21st Century* (see Chapter 1, "Lyric Oil").
6. The guide's point about wildlife should be expanded to include predators such as wolves and coyotes, which may be less willing than deer to incorporate disturbed areas into their habitat. A study of predator–prey interactions in Canadian National Parks shows that "anything more than 18 hikers a day on a trail was enough to affect the ecosystem by scaring predators from their prey" (Wingrove, "Nature"). On the other hand, research shows increased wolf–caribou predation in boreal forest that is fragmented by

energy exploration (Hebblewhite 76), and increasing white tail deer populations follow decreasing caribou populations as white tail deer forage on young trees, shrubs, and bushes while caribou eat lichen that live on trees in mature forests (Kopecky 40).

7. These two narratives can be understood as competing versions of what "community" means in Fort McMurray. For various sociological perspectives on this question, see the special issue of *Canadian Journal of Sociology* edited by Sara Dorow and Sara O'Shaugnessy.
8. The short story "Back" by Jennifer Quist includes the line, "I guess dying's one way to get out of town. That's the bright side of the whole thing" (4) before going on to address the narrator's surprise that there are cemeteries in Fort McMurray and realization that "some people just don't come from nowhere but here" (4).
9. In *The Energy of Slaves: Oil and the New Servitude*, Andrew Nikiforuk writes, "North Americans now live on credit to support their own energy slaves and to buy largely unnecessary goods created by other energy slaves" (72). It is not hard to see Donny's position in *Simple Simon* as such a form of slavery. See Chapter 6 for more on this idea.
10. Indeed, an industry representative at the 2006–2007 Oil Sands Consultation hearings makes this point explicit: "The reason I say we cannot simply give up on continuing use of fossil fuels, it's that sustainability is not just about sustainability of our environment, it's sustainability of our economy and our way of life, which like it or not requires energy in increasing quantities in the next fifty to a hundred years as driven by the developing economies" (qtd. in Gismondi and Davidson 135).
11. Žižek has, in characteristic fashion, responded to Confucius's point by arguing that "his teaching on *zhengming*, the 'rectification of names'...is itself a symptomatic misnomer: what needs to be rectified are the acts—which should be made to correspond to their names" (*Living* 13). While in general I agree with Žižek's point, in the case of bitumen at least, our acts are affected by what we choose to call them as well.
12. Indeed, this framing serves to ensure that development is a foundational good that cannot be questioned. Gismondi and Davidson provide this example: "as preface to the provincial Oil Sands Consultations, a Terms of Reference was developed to guide the process, which states, once again, that 'the oil sands area is key to Alberta and Canada's energy security, with $80 billion worth of projects already announced.' This and similar prefaces to the proceedings were very effective in establishing the parameters of debate. In short, development will not be subject to questioning" (159). Larry Pratt writes of how the view of Albertans as conservative is bizarre given their relationship to growth: "Alberta's dominant ideological outlook is characterized by an upbeat emphasis on change and the future, a pronounced tendency toward intervention in the economy, expanding bureaucracy and budgets, and a fixed belief in the absolute necessity for industrialization" (98). While some of these characteristics may have changed since Pratt was writing in 1976, it is still broadly true that Albertans are mis-characterized as "conservative." C.A. Bowers asks of conservatives what it is they want to *conserve* (*Mindful* xi): when

this question is asked of contemporary Albertans, at least as represented by our government, the answer appears to be "growth."

13. The Royal Society of Canada report, *Environmental and Health Impacts of Canada's Oil Sands Industry*, states, "Contrary to popular belief, the legislation and regulations do not require reclamation to boreal forest ecosystems, although this expectation has evolved over time with applications to operate and subsequent approvals" and "Reclaimed conditions will resemble and function as natural landscapes, provided that the legislated requirements are fully implemented, but reclaimed conditions will not be identical to the pre-disturbance state" (Gosselin et al. 10). Further, if the oil industry's past actions are any indication, even this promise of simulacral reclamation should be questioned. Bayne and Dale point out that "only 8.2 percent of conventional seismic lines in boreal Alberta reached 50 percent cover of woody vegetation after 35 years...A substantial percentage (about 25 percent) of seismic lines never return to forest as they transition into roads, pipelines, or buildings" and "In northeast Alberta about 36 percent of 15-year-old well pads in upland forest have fewer than 1,000 tree stems per hectare, 24 percent have 1,000–6,000 stems per hectare, and 39 percent have more than 6,000...A typical regenerating cut block is expected to have about 10,000 stems per hectare" (104).
14. Threats to the boreal caribou population in the bituminous region have been met with federal Environment Minister Peter Kent's attempt to "distinguish between protections for species that are at greater risk versus those that are only at risk in some areas" (De Souza A13) and plans to create a 1,500 km^2 "exclosure" to "protect caribou from predators such as wolves" (Canadian Press, "Caribou"). What this might mean for the wolf population is unclear.
15. The Royal Society of Canada found "the potential for successful reclamation of wetlands, particularly peatland, has not been well demonstrated in the research to date although studies elsewhere indicate more feasibility than generally believed." Furthermore, "The reclamation of tailings ponds and the EPL approach raise many questions about feasibility because no tailings pond has yet been fully reclaimed" (Gosselin et al. 11).
16. With the shift from the Energy Resources Conservation Board to the Alberta Energy Regulator in June 2013, the definition of *stakeholders* has been narrowed (see Thibault and Canadian Press, "Alberta Energy Regulator"). In June 2014 two First Nations filed suit against the Alberta Energy Regulator with the Alberta Court of Appeal, arguing that they had unfairly been denied standing in hearings regarding Canadian Natural Resources Ltd.'s Kirby expansion (Weber).
17. Allan Stoekl comments on this eventuality: "At some point the hubris of Man...runs up against a profound barrier: there is an enormous amount of energy *that is not servile.* It spirals out of celestial bodies, it blows away in the wind, it courses uselessly through our bodies" (*Bataille's Peak* 224), and it will remain locked in a matrix of sand, water, and clay—too deep, too diluted, or too dangerous to serve human ends.

18. Since this was written, Alison Redford abruptly resigned as premier of Alberta. Redford's resignation has been attributed to her personal sense of entitlement and her autocratic style of leadership, which may have threatened a caucus revolt, as well as to a misogynistic double-standard on the part of the Progressive Conservative Party leadership (see Laurie Adkin, "The End of Alison," *Alberta Views*, May 2014). Regardless of what went on behind the scenes, fear of losing the next election, under Redford's leadership, to the even more conservative and pro-energy Wildrose Party was a major factor.
19. It will be interesting to see if the NDP victory is a version of "only Nixon could go to China" (in this case, only the NDP can sell the oil sands as a "sustainable development" and justify new pipeline construction). *Edmonton Journal* business columnist Gary Lamphier has written of the NDP "with a bit of luck and some foresight, it might be able to salvage Alberta's oil industry" ("Time to Act" B3) and, following the appointments of Alberta Treasury Branch's David Mowat to oversee the royalty review and University of Alberta economist Andrew Leach to advise on the province's climate change strategy, "The new NDP government offered further evidence...that it plans to track toward the middle ground, and not lean too far left" ("Notley's NDP" C1). By comparison, consider Žižek's reflections on the 2008 election of Barack Obama in the United States in "Use Your Illusions," where he compares that event with Kant's thoughts on the possibility of ethical progress: "we can discern signs which indicate that progress is possible. The French Revolution was such a sign, pointing towards the possibility of freedom: the previously unthinkable happened, a whole people fearlessly asserted their freedom and equality" ("Use"). Indeed, despite consistent leads in polls leading up to the election, most commentators doubted the possibility of an NDP majority, and the result of the election was greeted with surprise by almost everyone. Like Obama's victory, the NDP win is significant "not only [because], against all odds, it really happened: [but also because] it demonstrated the possibility of such a thing happening. The same goes for all great historical ruptures—think of the fall of the Berlin Wall" (Žižek, "Use"). Will the NDP win enable Alberta to lead the transition to a post-carbon economy? Only time will tell, of course, but, as Žižek argues, "The true battle will begin now, after the victory: the battle for what this victory will effectively mean" ("Use").
20. Larry Pratt, in his seminal *The Tar Sands: Syncrude and the Politics of Oil*, argues that Alberta has even more faith in growth than most jurisdictions: "Contemporary Alberta sometimes seems obsessed with growth. If the rest of Canada is beguiled by, but also more than a little cynical about the flawed promise of limitless economic development, among Albertans the doctrine of open-ended progress through individual initiative is virtually an article of faith" (97). I would argue, that since Pratt wrote those words, this neoliberal faith has found many powerful converts around the world.
21. See Leslie Shiell and Colin Busby's report for the C.D. Howe Institute, "Greater Saving Required: How Alberta Can Achieve Fiscal Sustainability from Its Resource Revenues," for an example of this sense of "sustainable development."

22. See LeMenager's analysis of her trip to Fort McMurray in *Living Oil*; of particular interest is the consideration of Fort McMurray as a place where "we can see the future" (159), as a fellow visitor told her. She argues that her experience of Fort McMurray was "hyper-mediated" because of the "high density of metaphor in the place, the constant slippage of verbal images away from conventional referents toward new, sometimes unfinished, concepts" (161). This slippage is characteristic of the way factual and counterfactual statements interact and of the rhetorical deployment of enthymemes (see Chapter 3).

1 Lyric Oil

1. "Overburden" is often used to describe everything that overlies bitumen deposits. Grant, Dyer, and Woynillowicz describe the process of preparing a bitumen mining site this way:

> Before mining can begin, the forest, wetlands, and mineral soil are cleared, drained, and removed. Rivers and streams are diverted and forests are clear cut, with merchantable timber being harvested and the remainder being piled and burned. In addition, the layers of wetland and muskeg...must be drained and excavated.
>
> The muskeg layer lies about 1–3 metres (m) thick above the overburden (the material that overlies the ore deposit). Prior to the muskeg layer's removal, it must be drained of its water content...The overburden is then mined with large shovels and moved by dump trucks to be placed in above ground waste dumps (called overburden dumps) or mined out pits. The overburden may also be compacted into large dykes, creating dams that will eventually contain tailings.
>
> With the boreal forest and overburden removed, the oil sands ore is exposed and can then be mined. (5)

2. The certification process for reclaimed land is long and complex and, some suggest, is still being developed. The Alberta government says that nearly five thousand hectares have been "permanently reclaimed" and are undergoing monitoring until "ecological trends are achieved"; then "the company can apply for reclamation certification" (see the Alberta Government's *Alberta Oil Sands* website, "Reclamation" page at www.oilsands.alberta.ca/reclamation.html). Still, as of June 2015, no mined area has yet been certified as reclaimed and returned to the Crown.

3. Despite my belief that these costs are incalculable, Leach offers some startling numbers: "The Pembina Institute would put that figure at a whopping $15 billion—a big red number in a province that claims to have no provincial debt. The government of Alberta puts the current reclamation liability for existing mining projects at $5.5 billion, while the total security deposit posted for these facilities is $920 million." He concludes that "Given the magnitude of the reclamation liabilities the province has assumed and

will continue to assume under the new program to accelerate oil sands development, Albertans should ask some tough questions" ("A False Hope"). See also Jeff Gailus, "Black Swans" in the October 2012 issue of *Alberta Views.*

4. Wapisiw Lookout is the name of a Suncor "reclamation" project, the former site of a tailings pond that has been "reclaimed." It is important to note that to "reclaim" this site, Suncor first pumped the tailings to a different holdings pond (see LeMenager 162 and www.whatyescando.suncor.com).
5. Currently, a joint venture between Sherritt International and the Ontario Teachers' Pension Plan, entitled "Carbon Development Project," for a coal mine and gasification plant at the same place as in *Far as the Eye Can See* is working its way through regulatory channels. Originally proposed in 2007, the project has been delayed, according to the company, pending "greater clarity on carbon dioxide regulations and we are able to better align our project with the timing of our potential customers" (Cotter). Information about the development can be found on the Alberta Environment and Parks website at http://esrd.alberta.ca/lands-forests/land-industrial/programs-and-services/environmental-assessment/documents/Sherritt_Dodds-Roundhill_PDD.pdf.
6. Part of the hagiography of Peter Lougheed following his death on September 13, 2012 has been to contrast his views with the practices of current governments (Stenson 18). This trope entered the discussion about bitumen development around comments he made before his death (Nikiforuk, *Tar Sands* 141, 150, 177); however, Lougheed's main contribution to the discussion, both as premier of Alberta (1971–1985), and subsequently, was to encourage Albertans to act as "owners" of the resource (Marsden 77). While such a stance can lead to Albertans getting more money from each barrel of oil produced, it will only change views on environmental stewardship and protection if it is also connected to an understanding of "owing" that is non-provisional (As an owner of the land [not just the "resource"], I have a responsibility for the condition of the land into the future and long after my death. If doing something to it means that I cannot ensure its ability to sustain future generations of "all my relations" [King ix], including my non-human relations, then I will not do that thing.)
7. In their analysis of the contributions by Albertans to the 2006–2007 Oil Sands Consultation hearings, Gismondi and Davidson conclude, "the message conveyed by these individuals does indicate the presence of a distinctive resistance frame with resonance on several levels—in terms of rights, responsibilities, virtues, and ecological consciousness. And it is being expressed by Albertans themselves, who have the unique power to challenge directly the endorsing state apparatus" (186).
8. At the 2012 Association for Literature, Environment, and Culture in Canada (ALECC) conference, Nicole Shukin connected Lauren Berlant's concept of "slow death" to the twenty-kilometre evacuation zone surrounding Tokyo Electric Power Company's damaged Fukushima Daiichi nuclear power plant ("Place"). When people like Anthony Ladouceur, councillor of the Athabasca Chipewyan First Nation, say, "The places where

we used to pick berries and find our medicines have been destroyed by rapid tar sands projects...Our people have lived here for thousands of years, but it is becoming increasingly difficult to continue to live off the land with industry expanding all around us" (CBC News, "Healing Walk"), we can see that this concept applies to Alberta's boreal region as well. People downstream from the mining operations have been complaining for years about high cancer rates in their communities. Chapter 2 deals with these issues in more detail.

9. These lawsuits focus on an alleged lack of government consultation when approving bitumen leases and a failure to ensure constitutionally enshrined treaty rights to continued hunting, fishing, and trapping on traditional Aboriginal lands. These rights, however, are framed, like Lougheed's defence of agricultural land in Wiebe's play, in terms of *uses* rather than wilderness. University of Calgary Law Professor Shaun Fluker is undertaking work on the inability of the legal system to recognize the value of wilderness, and the language used in treaties provides another example of this. See Chapter 4 for more discussion of treaties as a possible site of resistance to bitumen extraction.

10. The Royal Society of Canada report concluded,

> As is commonly the case in boom-town style developments, there are major negative effects on community health due to several simultaneous pressures on individuals, families, and infrastructure. These include general price inflation, extreme housing shortages, labour shortages in most sectors, family stress, drug and alcohol abuse, increased crime, and other social negative impacts related to inadequate public health and municipal services. Community health status in the oil sands region is lower than the provincial average, which is dominated by Calgary and Edmonton data, but also poorer than that in the more comparable Peace Country region which has had its own, less intense, economic boom. (Gosselin et al. 12)

11. Since publishing those comments, Levant has written the book *Ethical Oil*, which led to the creation of the website and advocacy organization Ethical Oil (www.ethicaloil.org). Both the book and the website engage in remarkable rhetorical gymnastics to disguise the base materialism of Levant's original claims: rather than everything having its price, Levant and Ethical Oil argue that the sacrifices required for bitumen extraction are inconsequential compared to those required for other forms of oil (see Chapters 2 and 3 for more on this view).

12. See also O'Shaugnessy and Krogman for a more complicated view of gender in relation to resource extraction.

13. The suggestion that Alberta is a dictatorship has some support as well; see Nikiforuk, "The First Law of Petropolitics" (*Tar Sands* 152–66).

14. Ducks, of course, are not the only species that suffers mortality as a result of bitumen extraction processes. Erin M. Bayne and Brenda C. Dale point out that "open pits where waste fluids from hydrocarbon production are stored pose a significant threat to birds in some situations," and they mention the deaths on Syncrude's Aurora pond before noting, "Whether waterfowl are the species most affected by open pits is debatable; the US Fish and Wildlife Service identified the remains of 172 avian species (predominantly songbirds) in sampled holding ponds" (98). As part of the study into bird deterrent systems, funded through the fine Syncrude paid after the ducks died from landing on its Aurora tailings pond in 2008, University of Alberta biologist Colleen Cassady St. Clair found that, despite deterrent systems in place, thousands of birds are still landing on the ponds. The observers recorded "20,540" birds landing on the ponds in one season, and "given the observers could see only about 10 per cent of the surface of the ponds from their perch, St. Clair extrapolated the data for the entire pond. That meant there were more than 200,000 bird landings across the ponds, some of which would be repeat landings, between April and the end of October 2012" (S. Pratt, "Air Cannons"). St. Clair found that only about 1 per cent of the birds who landed on the ponds became coated with bitumen and died. However, "environmental regulations require the companies to keep birds off the ponds because they contain toxic substances...[so] 'It's still unacceptable to have birds landing in bitumen, so you have to get that bitumen off the pond,'" St. Clair said (S. Pratt, "Air Cannons").
15. The varied meanings of the word *science* are significant here. Pertinent definitions include "the state or fact of knowing"; "knowledge acquired by study"; "A branch of study which is concerned either with a connected body of demonstrated truths or with observed facts systematically classified and more or less colligated by being brought under general laws, and which includes trustworthy methods for the discovery of new truth within its own domain"; "In modern use, often treated as synonymous with 'Natural and Physical Science', and thus restricted to those branches of study that relate to the phenomena of the material universe and their laws...This is now the dominant sense in ordinary use." None of these definitions gets at Rowe's use of the term—the sense in which scientific knowledge is linked to technological application and involved in changing or attempting to control the world. Nor do they get at an important sense in which scientists often use the term—as connected to the scientific method, and, thus, always part of a process of doubt, critique, and reassessment of knowledge. In addition to these meanings, Timothy Clark writes, "'the scientific' so often functions as a kind of ideology, restricting important decisions to a culture of approved experts. Reference to science often serves to disguise issues of moral and political contestation under blunt assertions of supposed fact" (149). These tensions in the term are at play throughout *Unsustainable Oil* in the way that people use the term for rhetorical effect, to show that "science" is on their side in debates over bitumen extraction.

16. It is worth noting in this regard, as Gismondi and Davidson do earlier in their book, that "even assuming there *are* technological remedies looming on the horizon that might be of service, technology does not appear out of thin air. It requires heavy investments of capital and resources, including energy...What is more, the science that enables new technologies also requires ever more resources for ever-smaller increments of new knowledge" (147). Nikiforuk echoes this point in a chapter on "Peak Science" in *The Energy of Slaves*. He notes, "Each new extreme oil technique, from oil sands to oil shale, seems to cost more, takes longer to develop, and typically exceeds predicted costs, only to deliver less volume and even fewer energy returns" (*Energy* 167).
17. Stephanie LeMenager recounts a similar story from her Suncor tour: her guide said, "We remove everything, clean up the oil spill which Mother Nature left us, and put everything back." She notes, "This equation of intensive mining with cleaning 'nature's mess' was offered without irony" (163).
18. This idea also resonates in Alberta's 2009 rebranding exercise, which replaced the slogan "The Alberta Advantage" with "Freedom to Create. Spirit to Achieve."
19. In *The Energy of Slaves*, Nikiforuk shows how this has, indeed, come to pass, and was occurring even as Kroetsch was writing: "As farming, once a renewable enterprise, morphed into a giant mining operation governed by fertilizers, pesticides and machines, it became impossible to sit down to a meal in many parts of the world without literally eating oil" (74). This complicates the slogan "No blood for oil" when we realize, in fact that blood *is* oil, and the full vampiric implications of petroculture start to become apparent (see Chapter 6 for more).
20. See Chapter 2 for more on the lawsuit resulting from these deaths.
21. Bertha refers the superintendent to "Isaiah chapter six" (189) as proof of her understanding of the angel's anatomy, but the reference also indicates that the angel is a seraph, a creature who declares "the whole earth is full of [God's] glory" (Isaiah 6.3). This glory, then, also includes all the relationships of which the bituminous sands are a part. My thanks to Agnes Kramer-Hamstra for pointing out this connection.
22. Karl Clark, writing in a 1929 report called *Commercial Development Feasible for Alberta's Bituminous Sands*, stated that decisions about when to develop bitumen deposits would be based on economic rather than technical considerations.
23. This view of compensation was expressed in stark terms during the consultations undertaken by the Joint Review Panel, which was mandated by the Minister of Environment and the National Energy Board to assess the environmental impacts of the proposal and review the application under existing legislation. Northern Gateway witness John Thompson stated, "Although a spill could have a big impact on the fishery...compensation and other opportunities—such as working on clean up crews—will ensure people don't lose any income" (James).
24. Chapman became executive director for Northern Initiatives with Enterprise Edmonton in 2013.

25. Chapter 6 undertakes a reading of Richard Van Camp's *Godless But Loyal to Heaven*, which also suggests that the land contains that which is, from today's modern-liberal perspective, unthinkable. By linking bitumen extraction to a zombie apocalypse, Van Camp presents a stark answer to the question posed by Wiebe's story, which was written during the infancy of the commercial bitumen industry, around the time Syncrude shipped its first barrel of oil in 1978. Wiebe is asking us to choose to sacrifice the land and all it contains or to choose something else. Van Camp shows the consequences of choosing to make those sacrifices.

2 Oil Sacrifices

1. One example of such an alternative is the lament, which, rather than justifying or rationalizing the loss, simply states it, and dwells in that loss. Colleen Thibaudeau's poem "All My Nephews Have Gone to the Tar Sands" simply articulates the difficulty of isolation from family:

> I'm wondering
> Nephews, is what I feel writing you neon letters
> The same feeling my grandma had
> When she would turn and look out the window,
> (Staring just above the frostline & geraniums)
> Her voice flat and inevitable: "They've gone to the West."
> All my nephews have gone to the Tar Sands.
> I find it difficult to write to them. (248)

2. Gismondi and Davidson also point to *National Geographic*: "Aboriginal peoples living downstream from tar sands operations have become increasingly alarmed, and mobilized, in response to deteriorating health conditions experienced in recent years. In the *National Geographic* Photo Essay on the Oil Sands is Emma Michael, whose story resonates with many Northern residents. Breast cancer survivor, she stands graveside in the cemetery in Fort Chipewyan, Canada. Her brother died of colon cancer 2 years ago, her sister died of a bone-blood cancer 6 months ago and her mother died of cancer 4 months ago. See http://ngm.nationalgeographic.com/2009/03/canadian-oil-sands/essick-photography" (183).
3. It is interesting that Levant, a "conservative," argues on "progressive" grounds: the mythology of Ethical Oil claims that when Levant was unable to persuade an audience of the benefits of bitumen extraction based on "conservative" grounds, he developed the concept of "ethical oil" on "progressive" grounds, imagining an audience of "liberals." While Levant and the Ethical Oil movement have been widely criticized, it is also undoubtedly the case that the concept has found public resonance and effectively countered some of the attacks on the bitumen industry. I think there is good reason to critique

Levant's "progressive" concept on conservative grounds, by asking what, precisely, is it intended to conserve? The answer, it seems to me, is "progress." Given that Canada is a modern, liberal, technological society, it is unsurprising, perhaps, that an argument from progress should be persuasive.

4. See Shaun Fluker's "*R v Syncrude Canada:* A Clash of Bitumen and Birds" where he concludes, "At some point we cannot have both habitat protection and unabated energy development—there are choices to be made. The enactment of 5.1 of the [Migratory Birds Convention Act] demonstrates a choice made by Canadians (including those who reside in Alberta) in favour of habitat protection. Reading down this provision in an attempt to accommodate both tar sands development and habitat protection makes a mockery of the law, and misleads all of us into thinking sustainable development is a priority in our society" (244). In 2014, Alberta biologist Kevin Timoney said that "if the governments don't require stronger deterrence measures, the federal Migratory Birds Act will become meaningless." He added, "No matter how you look at it, we have a good law that is not being enforced" (qtd. in S. Pratt, "Eyes of the World" A4).
5. See Nikiforuk, *The Energy of Slaves.*
6. Devin Ayotte argued at the 2013 Petrocultures conference that if industry was committed to worker safety, instead of implementing ever more complicated safety protocols, they would require equipment to be engineered differently.
7. With this formulation I want to suggest simultaneously eutopia as "the good place" and utopia as "nowhere" or a "non-place."
8. As one speaker at the 2006–2007 Oil Sands Consultations, designed by Alberta MLAS to allow Albertans to share their views on bitumen development, asked the panel, "tell me how this market, who's [*sic*] only determination of success is profit, is going to ensure that no more people will be left homeless, that oil sands will not be our single largest contributor to greenhouse gas emissions, that no more children will be kept indoors at recess, or that no more people will die" (Oil Sands Consultations, Edmonton, 3 April 2007 qtd. in Gismondi and Davidson 87). See also Brenda Longfellow's film *Dead Ducks,* which imagines the "Executive Training Centre" of the public relations firm "Oxbow and Associates" where, as the duck deaths are unfolding on the Aurora tailings pond, Peter Keleghan instructs his trainees that, when dealing with a public relations disaster, it is important "not to dwell on the past." Later he describes the art and science of the perfect apology, the goal of which is to be "forgiven": "don't ever offer excuses," he says, and "it is always about *them* and how they feel." Later in the film we see these strategies played out with a very smug looking Syncrude CEO describing the company's new avian-deterrent system and the negative press being countered by letters to the editor comparing the duck deaths to the number of birds killed by skyscrapers, domestic cats, and wind turbines. The film counters this move to minimize the suffering of the ducks by quantifying and relativizing, with the kidnapping of the surviving ducks from the Edmonton

Wildlife Rehabilitation Centre to prevent their story from being used as a part of a PR campaign.

9. Given such a situation of infinite responsibility, Angus writes similarly, "The inheritors of dispossession know that survival means implication in the mechanisms of continuing dispossession. Justice must be the rebellion against this implication in the mechanisms of continuing dispossession" ("Continuing Dispossession" 8).
10. Funding a program in Wildlife Management at Keyano College and funding a University of Alberta study on bird migration patterns and the effectiveness of bird deterrent systems suggests that through improved scientific study and, through the application of that study, improved technological control of nature, our historical striving will be, if not realized, at least advanced. Similarly, the conditions imposed by the Joint Review Panel into the Northern Gateway Project "require Northern Gateway to follow through on all of its commitments, including...funding a marine research chair at one or more British Columbia universities to examine knowledge gaps about ecosystems in the Pacific north coastal region and to develop research programs in those areas" (NEB 48).
11. For more on the details of this process, and the role of the Alberta government in *discouraging* investment in bitumen in order to protect the conventional oil industry from competition, see Gismondi and Davidson 58–60.
12. It is significant in this respect that industry makes use of the symbolism of bison, the largest North American land mammal and icon of "The West," an animal that was driven to the brink of extinction by "progress" and was completely extirpated from its natural habitat (see Tasha Hubbard's "'The Buffaloes are Gone' or 'Return: Buffalo'?: The Relationship of the Buffalo to Indigenous Creative Expression") and is now recovering rather than an animal currently threatened by industry such as the woodland caribou. See Chapters 3 and 5 for more discussion of boreal caribou. Hubbard quotes Lame Deer as saying that for Plains indigenous people, "The buffalo was part of us, his flesh and blood being absorbed by us until it became our own flesh and blood. Our clothing, our tipis, everything we needed for life came from the buffalo's body. It was hard to say where the animal ended and the man began" (68). While indigenous people in the boreal region had a different relationship with wood bison, whatever role Syncrude's herd is playing, it is neither restoring a pre-colonial relationship nor moving to a postcolonial one.
13. Dene Chief Allan Adam makes this point in the film *Tipping Point: The Age of the Oil Sands.*
14. Gismondi and Davidson describe Ells's *Recollections* this way: "From opening poem to closing words, Ells sketches a manly frontier where skilled 'white' engineers supervised urban tenderfoots, with the assistance of shiftless Native labor. Ells' memoir includes fascinating images and descriptions of quarrying techniques, use of explosives, power shovels, and shalers in early mining activities, alongside discussions of various paving tests and results of trials at separation of the oil from the sands" (46).

15. For another example, as part of Syncrude's advertising campaign, see "Steve Gaudet, B.Sc., Syncrude Canada," www.youtube.com/watch?v=iXtHFPGXjlY, uploaded 11 June 2010.

3 Impossible Choices

1. A rant by Lisa Boutilier, posted to her Facebook page, and reprinted by Verna Murphy in the *Cape Breton Post* makes a similar point: "If you hate Fort McMurray—leave. Don't come and work here long enough to get your stamps that allow you to sit on your arse the rest of the year and then have the nerve to complain about coming back out for another 16 weeks. Don't take the paycheque from here to pay your mortgage, make your car payments, and take your family on vacation and then bitch about how much you hate this place...Personally I think that if the community I lived in couldn't provide for me and my family, then *that* is the community I would hold a grudge against. Not the one paying my bills. I always say nobody hates Fort Mac on payday, but if you do—go home. Seems simple to me" (qtd. in V. Murphy 24).
2. Sara Dorow and Goze Dogu see at least three Fort McMurrays: that of the long-timers working to build community, that of the oil workers whose community resides elsewhere, and that of the oil industry professional for whom "many of the roots they had selectively put down in Fort McMurray were offshoots of their portfolios of degrees, networking skills, globalized identities, and personal values. Fort McMurray seemed to offer enough of the right things right now for creating a life" (286).
3. The prominence of the airport in the maps produced by Dorow and Dogu's interviewees emphasizes this point.
4. In August 2013, another controversy arose following the use of a Travel Alberta slogan, "Remember to Breathe," in a fundraising campaign for a film criticizing bitumen extraction, "Welcome to Fort McMoney—Remember to Breathe" (Stephenson).
5. As the Sierra Club's protest signs against the Keystone XL Pipeline have it, "There is no Planet B"; or as Maxine Burkett reported at the 2013 ASLE conference 2013, many inhabitants of island nations threatened by rising ocean levels have declared their intention to drown where their ancestors are buried rather than leave their homes.
6. He "knows engines."
7. "Because they pay attention to what you buy."
8. Steve Fuller notes that, "The bullshitter normally refers to things whose *prima facie* plausibility immunizes the hearer against checking their actual validity. The implication is that the proof is simply 'out there' waiting [to] be found" (249). Is it true that increasing trade will ensure financial security? What constitutes "financial security"? How does "financial security" relate to other values with which it may compete or co-operate?
9. For a striking juxtaposition of different "facts" about bitumen extraction, see Gailus, *Little Black Lies* 18–21.

10. Žižek has also argued that we should "reject the underlying premise of Harry Frankfurt's critical analysis of bullshit: ideology is precisely what remains when we make the gesture of 'cutting the bullshit'" (*Living* 3). Žižek's analysis here is unsatisfactory: while ideology may be what remains when one "cuts the bullshit," surely ideology also operates through bullshit in Frankfurt's sense—when one speaks without knowing what one is talking about what ends up speaking is not the individual but the dominant discourse of the culture, ideology. Regardless, my claim here is not that we should "cut the bullshit," as if that were possible in some absolute sense, but that the counterfactual move to "matters of concern" can help us get beyond the truth/bullshit binary.

11. For more on the mediated nature of the public's understanding of the bituminous sands, see Gismondi and Davidson (29) and LeMenager (161).

12. In closing arguments at the Northern Gateway hearings, Enbridge lawyer Richard Neufeld warned of an "economic catastrophe of unprecedented proportion" (Cattaneo, "Northern Gateway") if Northern Gateway is rejected. He also responded to criticisms that Enbridge has not supplied all necessary information about the project by saying, "No amount of additional...information would persuade any member of the tar sands campaign to support a pipeline such as this. They are never going to say that enough information has been provided" (Cattaneo, "Northern Gateway"). Now that the hearings have concluded, the focus of debate has shifted to the 209 conditions placed on the project by the NEB's Joint Review Panel (see Chapter 5).

13. Bill Van Heyst, associate professor of environmental engineering at the University of Guelph, has warned that cuts included in the federal Conservative government's 2012 omnibus budget Bill C-38 will degrade the quality of environmental monitoring even further (De Souza A13).

14. For a concise history of environmental monitoring in the region, see Gailus, *Little Black Lies* 21–41. Schindler has also argued that "the role of science in environmental policy has continued to weaken. There has been a terrible devitalization of science and scientists in both federal and provincial departments. The governments have made their scientists the tools of politicians, their results and opinions twisted if necessary for political gain" ("Climate Unchanged" 19).

15. By contrast, the documentary *Tipping Point: The Age of the Oil Sands* suggests that Schindler's research will prove to be a watershed moment for the bitumen industry, though this has not yet happened in my opinion.

16. Regarding the sustainability of CCS projects for bitumen operations, see Grant, Dyer, and Woynillowicz, "Fact or Fiction: Oil Sands Reclamation."

17. A third move involves, as Gismondi and Davidson note, dismissing criticisms of government and industry as being anti-development, and, therefore, outside of the frame of legitimate discourse (130). Joe Oliver uses this move, stating that those who oppose development are "inimical to Canadian national interests" (Cattaneo, "A Battle").

18. Enbridge lawyer Richard Neufeld makes the same point: "Tradeoffs are a fact of life... That does not mean that any person or community or region must be marginalized...All it means is that in determining public interest we need to seek balance that respects local interests, plans that deliver benefits to local communities, while still ensuring that the projects that are needed for this country will proceed" (Cattaneo, "Northern Gateway"). Former chair of the Oil Sands Developers Group Ken Chapman touts the "triple bottom line," which attempts to account for economic, social, and environmental aspects of a company's operations (Chapman, "The Next North").
19. Oliver claims to be "bothered by their [environmentalists'] misrepresentation, hyperbole, statements he says they know aren't true" (Cattaneo, "A Battle").
20. On June 17, 2013 the ERCB was replaced by the Alberta Energy Regulator (AER), changing its structure from an independent agency of the Alberta government to a corporation operating at arm's length from the government. While I am not in a position to go into the implications of this change here, it is worth noting the different language, as the AER is responsible for overseeing the transition from the former Energy Resources Conservation Act to the new Responsible Energy Development Act (Jamieson, Olynyk, and Ference): rather than "conserving" energy resources, we are now "responsibly developing" them. It will also "ultimately [in 2014] combine the regulatory powers of the former Energy Resources Conservation Board (ERCB) and Alberta Environment and Sustainable Resource Development (AESRD) into one body, the AER" (Thibault). Alberta Energy's 2012/2013 annual report describes the transition as one that will "provide industry with an effective, transparent and streamlined process to approve new projects and engage with stakeholders" (28). Though it is referred to as a "regulator," its purpose is, first of all, to approve projects.
21. See Chapter 1 for more on "lyric thinking."
22. "And that's interest of the piece. *Fort Mac* is not an ecological fable that works to expose the bituminous sands. This is not a political play, claiming a clear and precise anti-capitalist position. This is simply a human piece—the man, the woman—who, through sincere devotion to a cult of money and prosperity, get caught in a whirlwind without the power to get out. The characters reveal, directly to the audience, fragments of their humanity bit by bit, humanity crumbling throughout the play."
23. "Kiki, a modern day Virgin Mary, prays, in vain. She is the only one to believe in something other than money."
24. We might read her death as a gift, which, in Stoekl's words, "escapes the intentions of its 'author.' What is important is gift-giving itself, and the good or bad (or selfish) intentions of the giver are virtually irrelevant. What counts, in other words, is *how* one spends" (*Bataille's Peak* 141). I take this to be quite different than the version of liberalism that also claims that "a fully self-conscious liberal should intentionally limit his altruistic readiness to sacrifice his own good for the good of others, aware that the most effective way to act for the common good is to follow one's private egotism" (Žižek, *Living* 36). In such a

situation there is still the possibility for such selfish behaviour to escape the intentions of the liberal, but the *how* here is clearly operating under "a certain model of utility associated both with the object and with the self" (Stoekl, *Bataille's Peak* 141). Even if, ultimately, "The left hand never really knows what the right is doing" (141), focusing on one's private egotism is an attempt to control both hands at once.

25. "If I die now my death will not have any meaning."
26. "I have found my reason for being. I am going to make people happy."
27. "Kiki is alone in recognizing the dangers of the bituminous sands, often invoking Mother Nature. But she especially represents the dangers associated with human greed."
28. See "The Art of Ecological Thinking," where Chisholm argues that Meloy "treats ecological upheaval as inexplicably affective, not as positive, measurable, factual, and fixable. Affect is key to Meloy's art of composing a vision of what is otherwise imperceptible and unthinkable" (572). In my reading of *Fort Mac*, the use of "Dame Nature" works to generate affect, despite the problematic associations of that phrase.
29. "If Mother Nature saw what was happening here in Fort McMurray, she would cry an ocean of tears. But now that I think of it...She does not need to see it—she feels it, she knows it. My God...She must feel it."
30. "I never understood the need of men to want to destroy what is beautiful. I just do not understand. Maybe they do not see the beauty that surrounds them—but I refuse to believe that these men cannot see beauty because that's what makes us human. Worse, I don't believe that these men are not human. They feel perhaps set apart—detached or superior to nature. Nature is only a force to subdue, to dominate, to exploit."
31. "What is it that you want from me?"
    "I want your innocence. Your light. Your beauty. Your benevolence."
32. "If you want me, take me."
    "Fuck...I cannot fuck the Holy Virgin, estie!"
33. "I'm right, am I not right?"
    "You're right."
    "I'm definitely right."
34. See Nancy, "The Unsacrificeable," and Chapter 2 for more discussion of this idea.
35. "Mother Nature is tired, too. She has provided us with all her bounty, and we thank her by profiting from her. We take advantage of her, abuse her...We violate her." Some ecofeminists criticize the anthropomorphism involved in the construction "Mother Nature." I do not have space to outline this debate here, but will simply suggest that, despite flaws and limitations, it still has capacity to help recontextualize an objectivist approach to environmental issues. Catriona Sandilands has described a goal for ecofeminism as being to "understand the ways in which nature and gender are wielded as discursive constructs, to investigate the ways in which the oppression of women and the domination of nature are imbricated in a whole host of destructive relations and practices, and to create an oppositional framework capable of addressing their interrelations" (xvi). Deployment of "Mother

Nature" as a discursive construct could serve the ends of domination in some contexts or operate as part of the oppositional framework in others. My contention is that in *Fort Mac* it operates to address the interrelations between the oppression of women and the domination of nature through the character of Kiki.

36. "to mark her departure, her suffering, her sacrifice"
37. "He throws the rose petals below the bridge."
38. For a bizarre attempt to rationalize such a move, see Ethical Oil spokesperson Kathryn Marshall on CBC's *Power and Politics* (January 11, 2012), an interview that started to go viral on YouTube before it was taken down, but is still available on cbc.ca: http://www.cbc.ca/player/News/Canada/Gateway+Pipeline/ID/2186009302/.

4 Irrational Oil

1. We also see this harmony in Kiki's life in Prescott's *Fort Mac*, discussed in the previous chapter.
2. Sourayan Mookerjea states, "Community becomes a coping strategy, a strategy of 'risk aversion,' bringing people together against the various faces of endured exploitation, of official indifference or neglect, if not contempt; or, community is aesthetically engineered and incited as some compensatory affective hype by community branding strategies... rolled out by industry and government public relations agencies" (234). Those who speak for "community," in either sense, work to establish a certain ethos.
3. In this sense we can see Ghandi, Martin Luther King Jr., and Nelson Mandela as parrhesiastes regardless of whatever other rhetorical strategies they may have employed.
4. While the contemporary North American equivalent of the king or tyrant (a politician or CEO) may indeed risk something by speaking in a certain way (may hurt their chances of re-election, may cause stock prices to fall), I have not encountered such speech from governing politicians while in office or current corporate executives. It is also significant that the privilege of those who speak against bitumen extraction is regularly used as a way of dismissing those criticisms. Michael Den Tandt, for example, responding to Archbishop Desmond Tutu's comments about bitumen states, "This growing anti-growth, anti-pipeline cant, driven by a global, ideological aristocracy seemingly divorced from reality and oblivious to contradiction, is doing real damage now to Canada's economic prospects, and putting ordinary people at greater risk by forcing crude oil onto railways" (A11).
5. For example, in considering the role of social movements in the bitumen debate, Gismondi and Davidson suggest skepticism about the potential for such movements to stop development. However, they claim, "Even if these organizations cannot stop the tar sands, or foster a transition *directly*, they can nonetheless play a critical *indirect* role, by introducing new ideas and a degree of scepticism into the public sphere" (177). However, increased scepticism is also an available response to any new ideas introduced by social

movement groups for those who would preserve the status quo (such scepticism is part of the sophistry exhibited by ethicaloil.org, for instance).

6. It is important to note that Innes's study is focused on the Cowessess First Nation of southeastern Saskatchewan and the role of Elder Brother in that community. He notes the work of Kristina Fagan, who criticizes the way "literary critics view trickster characters in such a generalized manner that it ignores cultural and geographical specificity" (25). I recognize that I am on uncertain terrain here, using a Greek concept (*parrhesia*) as read by a French theorist in conjunction with a Native figure. However, I trust Innes's declaration, that "the underlying role of all trickster stories in North America" (29), following Franchot Ballinger, is to "instruct and act as societal control by dramatizing community values and behavioral limits" (28). While the particular limits may vary with cultural and geographical specifics, and the methods certainly vary, the act of re-presenting such limits for a community is shared by both trickster figures and *parrhesiastes*.

7. An example of this strategy was seen following Greenpeace direct actions at Shell's Albian Mine and Fort Saskatchewan upgrader in 2009. The Alberta government's response, labelling the activists as "terrorists," quickly detracted from the message Greenpeace was intending to send, which relied, for its impact, upon the personal risk the activists endured (Arsenault).

8. Steve Fuller asks, "what if it were made easier to assert and challenge knowledge claims without having to undergo the personal transformation of, say, doctoral training? In the absence of such institutionalized immunity to bullshit, the result would be a Sophist's paradise. Truth would be decided on the day by whoever happens to have the strongest argument or survives some mutually agreed test" (246). Further, as his discussion of Wittgenstein shows, it is not always possible to tell when someone is a bullshitter and when she or he is a bullshit detector and this distinction may shift depending on the context (248).

9. At the 2013 meeting of the Association for the Study of Literature and the Environment, at a roundtable titled "Building the Environmental Humanities," Paul Outka argued, "militaristic and technocratic discourse assumes that its language is the language of the real, and then says to the humanities 'say something in my language' and when humanists can't, don't, or won't do that, then the former define whatever the latter say as 'bullshit'" (Outka). The same is true, I think, in the bitumen debate and, as in the defence of the humanities, while such dismissiveness happens on both sides, one side is more powerful than the other. Humanists may claim that technocratic discourse is bullshit in the same way that technocrats may call humanistic discourse bullshit, but technocrats control university budgets in ways that humanists don't; similarly, environmentalists may

call industrial and governmental technocrats' discourse bullshit in the same way that those technocrats may dismiss environmental discourse, but the technocrats have more influence on policy than environmentalists do.

10. This response parallels that of the federal government to opposition to Enbridge's Northern Gateway Project (see Chapter 3 and, especially, the discussion of the editorial in the *Edmonton Journal* on January 10, 2012).

5 Pipeline Facts, Poetic Counterfacts

1. Nor, by any means, are these two collections the only artistic responses to the pipelines proposal. For more artistic responses, visit *People on the Pipeline* at www.peopleonthepipeline.com/pipeline-directory/.
2. The oil on the train that derailed in Lac Mégantic was not derived from bitumen; it was light crude that originated in North Dakota's Bakken formation. Nevertheless, the Lac Mégantic derailment brings home the dangers of transporting petroleum products by rail.
3. As a letter to the prime minister from three hundred natural and social scientists states, instead of

> offering a balanced consideration, the panel justifies its recommendation of the project by summarizing the panel's understanding of environmental burdens in five short paragraphs and judging that these adverse environmental outcomes were outweighed by the potential social and economic benefits. Without a rationale for why the expected benefits justify the risks (e.g., why must an environmental effect be certain and/or permanently widespread to outweigh economic benefits that themselves are subject to some uncertainty?), any ruling of overall public interest is unsupportable. (Chan et al. 1)

4. Northern Gateway made the same point in submissions to the Joint Review Panel (NEB 59).
5. Several of the poems in *The Enpipe Line* take this strategy even further by using found poetry, often promotional material for the pipelines, repurposed within the collection as opposition.
6. Darin Barney notes, "Recently, former Premier of New Brunswick and now Deputy Chair of TD Bank Frank McKenna invoked nationalist rhetoric in calling for an east–west oil pipeline to carry diluted bitumen from the Alberta oil sands to the Irving refinery in Saint John. Comparing the prospect to the building of the CPR, McKenna suggested that, 'in a country where gravitational forces often move north and south,' a latitudinal oil pipeline could 'knit the country together both symbolically and economically.'"
7. Despite the federal government's approval of the project on June 17, 2014, pending the JRP's 209 conditions, the future of the project remains unclear at the time of this writing

given Aboriginal opposition and the need for the BC government to issue construction permits. The pipelines promise to be a wedge issue in the 2015 federal election, with both the Liberals and NDP promising to overturn the Conservatives' decision, if elected.

8. The September 14, 2013 edition of the *Edmonton Journal* included a political cartoon triptych in which a woman is depicted in exactly the same pose in each of the three panels, as she responds to questions about oil transportation: "Pipeline?" asks the first. "Nope. It'll leak and pollute the land," says the woman. "Tankers? Trains?" asks the second. "No, too unsafe," states the woman. "How do we deliver your oil, then?" the third inquires. "That's not *my* problem," the woman concludes (Lamontague A22). While such denial of responsibility may be a common stance—"I want cheap energy, but I don't want to think about where it comes from"—exposing the inadequacy of that stance need not lead to the equally simplistic dismissal of criticism—"Your lifestyle requires energy: therefore you can't be critical of the energy industry." We are all implicated in the fossil fuel system; that does not mean we must defend the status quo.

9. The Joint Review Panel Report quotes Northern Gateway as saying that "the economic benefits over 30 years would include...907,000 person years of employment" (NEB 31). The direct employment provided during the three and a half years of construction, according to Northern Gateway, would be "13,870 person years" (NEB 31). The 62,700 person years of construction employment includes both direct employment (digging the trench, laying the pipe, building the terminal, etc.) and indirect employment (those involved in producing the steel for the pipe or producing the energy to power the steel mill, for example). It is also worth noting that "person years" of employment is quite different from "jobs"; a "person year" of employment is an accounting term for the amount of work done by one person in a year consisting of a standard number of work days. Thus, 62,700 person years of employment could be 62,700 people working for one year, approximately 10,000 jobs lasting six years, and so on, with various permutations and combinations.

10. It is worth noting, here, that the company uses the singular "pipeline" in their ad and in various communications, when, in fact, the project proposes two pipelines: one to carry diluted bitumen from Bruderheim, Alberta, to Kitimat, British Columbia, and another running the opposite direction to carry condensate, which is used to thin bitumen so that it can be transported by pipeline.

11. Or, more accurately, I am uncertain of the *rhetorical exigence* of such interventions. For Lloyd Bitzer, an exigence "is an imperfection marked by urgency" and "it is rhetorical if it can be modified with discourse" (qtd. in Gorrell 395), so the question is not whether there is an imperfection marked by urgency—I think any reasonable observer of the development of Canada's bituminous sands would recognize that there are many such imperfections and some urgency in addressing them—it is whether it can be modified with discourse. To date, all such attempts have been neutralized by the hegemonic discourse of modern techno-liberalism.

12. The relationship between the hope offered by such upward counterfactual visions of the future and the hope offered by "locative thought" (Angus, "Continuing Dispossessions") requires more exploration than I can engage in here. Suffice it to say that the temptation "to say that the plurality of stories of dispossession can be totalized into a universal concept of not-dispossession, emancipation through inhabitation, or justice" is, in my view, quite a different hope than the notion that application of facts through science and technology will, at some point in the future, balance the ongoing dispossessions required in that application with desires for inhabitation.
13. Gismondi and Davidson follow Bill Freudenberg in noting that

    environmental degradation can persist due to a rather masterful "double diversion." This entails, first, diversion of access to resources and waste sinks into the hands of a privileged few, combined with, second, the diversion of *attention* in discourse away from the resulting disproportionalities, allowing for the perpetuation of certain privileged accounts...Textual and visual information can channel public understandings, often in ways that are not immediately perceptible, and by means of discursive activities that are exercised in plain view, not behind closed doors, a form of showmanship likened to magicianship (Alario and Freudenburg 2006; Freudenburg and Alario 2007). (26)

    Maria Lavis suggests this type of charlatanism in her poem "Lingerers On": "Look, life is good / look at our facts and figures that tell a different story.../ over here, not over there,/ hey! Stop looking over there!"
14. I see Kant's idea of "balance" here as different from Collyer's and Neufeld's, discussed in Chapter 3, in that Kant's is embodied in a particular speaker whose very life depends on maintaining this balance (this would also be one way of understanding sustainability) but is willing to risk that life to speak (as a parrhesiastes), while Collyer's and Neufeld's is abstracted into the realm of discourse: the "trade-offs" they consider need to be "reasonable" and "practical" and "improve our engagement and communication with customers"; the actions necessary to the tightrope walker, however, are not subject to trade-offs or negotiations; each step must be taken exactly if balance is to be maintained.
15. Stephen Maher quotes Barack Obama from an April 2013 fundraiser making a similar point:

    If you haven't seen a raise in a decade; if your house is still 25,000, 30,000 dollars under water; if you're just happy that you've still got that factory job that is powered by cheap energy; if every time you fill up your old car because you can't afford to buy a new one, and you certainly can't afford to buy a Prius, you're spending 40 bucks that you don't have, which means that you may not be able to

> save for retirement. You may be concerned about the temperature of the planet, but it's probably not your No. 1 concern. (A18)

Undoubtedly, jobs are important, but they are not the only thing that is important. Jobs are good, but they are not the only, or highest, good. George Grant writes,

> In human life there must always be place for love of the good and love of one's own. Love of the good is man's [*sic*] highest end, but it is of the nature of things that we come to know and to love what is good by first meeting it in that which is our own...In many parts of our lives the two loves need never be in conflict. In loving our friends we are also loving the good. But sometimes the conflict becomes open. An obvious case in our era is those Germans who had to oppose their own country in the name of the good. I have known many noble Jewish and Christian Germans who were torn apart because no country but Germany could really be their own, yet they could no longer love it because of their love of the good. (*Empire*, 73)

In a similar fashion, one's own job may come into conflict with the good (see Matthew Henderson's *The Lease* for a visceral depiction of such conflict involved in oil labour), or, indeed, one's civilization, in which case individuals may feel themselves torn apart, they may deny the conflict (and rationalize the consequences of that civilization as actually being the same as, or bringing about, the good), or they may search for alternatives to affirm. It's also worth noting here that Rex Murphy has received speaker's fees from the oil industry (see Mitrovica), which may colour his pronouncements about its value, suggesting that his rhetoric is more that of the magician than the tightrope dancer.

16. It is also worth noting that the Joint Oil Sands Monitoring Program "doesn't monitor groundwater around in situ operations [which will account for 80 per cent of recoverable reserves]...but focuses only on the minable oil sands" (Kopecky 40). Arno Kopecky quotes Bill Donahue with Water Matters regarding the bitumen leak at CNRL's Primrose Operations near Cold Lake, a high pressure cyclic steam stimulation operation: "the CNRL spill was discovered by someone just walking through the bush who noticed this bitumen bubbling out of the ground. It wasn't because of a science monitoring program" (40; see also Timoney and Lee).
17. Gismondi and Davidson write of such stories, "Fortunately for state agents facing such a bind, the direct *indications* of environmental degradation are often obscure and are dependent upon scientific interpretation, creating the conditions for what Rothstein (2007) calls the Professional Model of the welfare state, in which state professionals invoke specialized knowledge in ways that appear impartial, privileging expert discourses while marginalizing others" (25).

18. As Clinton N. Westman puts it, "This tendency to conduct research after projects have already gone ahead is surely one of the most pernicious fantasies wrought by the sorcerer, petro-capitalism, and its genie, petro-politics" (216).
19. The Government of Alberta's wetlands policy announced in September 2013 is an extreme example of this: the policy, eight years in the making, will not take effect for two more years and exempts all 195 currently operating or approved bitumen projects (S. Pratt, "Policy").
20. As Steve Fuller argues, in "Just Bullshit," "What sells is ultimately not intrinsic to the product but one's idea of the product, which advertising invites the consumer to form for herself" (244). Natural Resources Canada spent $9 million in 2013 to promote the energy sector in its "Responsible Resource Development campaign" (Canadian Press, "Government") and a total of $40 million during 2013–2014 in Canada and internationally (Canadian Press, "Oil and Gas Ad Campaign").
21. It is also noteworthy that the federal government is "downgrading the protection of humpback whales off the coast of B.C. under the Species at Risk Act" from "threatened" to "species of special concern," given that one of the Northern Gateway Project's pipelines would end with a "tanker shipping route that overlaps with...'critical habitat' for the whale" (Chung).
22. Populations biologist Stan Boutin states that government "has never restricted industrial development or activity. The linear footprint is growing rather than declining. Government and industry have a very clear plan of how much they want to expand, and good for them. But there's no corresponding plan in place for how they're going to manage cumulative effects" (Kopecky 40).
23. Fuller notes that, "When venturing into terrain yet to be colonized by a recognized expertise, bullshitters assert authoritatively rather than remain silent" (245).
24. Parrhesia is dangerous to the listener (it threatens our sense of security by making us recognize the precariousness of our position) and to the speaker (the listener on being told of the danger of his/her position may respond with denial and "shoot the messenger").

6 Oil Desires

1. Žižek makes the point that

> nowhere are the new forms of apartheid more palpable than in the wealthy Middle Eastern oil states—Kuwait, Saudi Arabia, Dubai. Hidden on the outskirts of the cities, often literally behind walls, are tens of thousands of "invisible" immigrant workers doing all the dirty work, from servicing to construction, separated from their families and refused all privileges. Invisible to those who visit Dubai for the glitz of the consumerist high-society paradise, immigrant workers are ringed off in filthy suburbs with no air conditioning. They are brought to Dubai from

> Bangladesh or the Philippines, lured by the promise of high wages; once in Dubai, their passports are taken, they are informed that the wages will be much lower than promised, and then have to work for years in extremely dangerous conditions just to pay off their initial debt (incurred through the expense of bringing them to Dubai). (*Living* x)

2. In 2011, Google's energy use was equivalent to that required to power two hundred thousand (presumably North American) homes. Google claims that its services actually make the world greener "because people use less energy as a result of the billions of operations carried out in Google data centers. Google says people should consider things like the amount of gasoline saved when someone conducts a Google search rather than, say, drives to the library" (Glanz). This seems to me, though, like it may be an example of the Khazzoom-Brookes Postulate, which suggests that "increased efficiency in the exploitation of a resource will lead over time to greater consumption, not less" (Nikiforuk, *Tar Sands* 121): we can make a lot more Google searches than trips to the library; the cost per search is lower, but the number of searches increases exponentially. Such a statement also seems to imply that library usage will decrease as Internet usage increases; while there have been moderate decreases in North American library visits in recent years (5.8 per cent in 2011, "Public Library Data"), we should remember that libraries are one of the only places with free, public computer and Internet access for those who otherwise could not afford it. For another take on the limitations of the Internet and social media to effect social change, see Shukin, "The Trouble with Twitter."
3. This took its most explicit form in 2005 when each citizen was given a four-hundred-dollar "prosperity bonus" by the Alberta government. The so-called Ralph bucks, named for then-premier Ralph Klein, though much derided by critics as an attempt to buy the electorate with their own money, could also be read as a populist attempt to include the population in the windfall profits the government was then reaping from the sale of oil leases and royalties.
4. Clinton N. Westman writes, "the logic of petro-capitalist extraction resembles the predatory logic of the Windigo, a dangerous (and rapidly growing) entity that has lost its humanity and proper sense of relatedness to others" (221).
5. See Chapters 3 and 4 for more on the characterizations of environmentalists as "terrorists" and "radicals."

Conclusion

1. In a head note, the *McMurray Musings* blogger, Theresa, provides a rationale for writing: "It's time for Fort McMurray residents to tell their own stories. For many years we allowed those from other places to tell our story, and we often found those stories did not reflect who we truly are. This blog is my attempt to tell my own story of life in this community, and to share my story with the world."

Works Cited

Acuña, Ricardo. "Ending the CEMA Sham." *Vue Weekly* 27 Aug. 2008. Web.

Adkin, Laurie. "The End of Alison." *Alberta Views* May 2014: 34–40. Print.

Alberta Energy. "Alberta's Oil Sands." Web. 8 Jan. 2013.

———. *Energy: Annual Report, 2012–2013*. 3 June 2013. PDF file.

———. *Energy: Business Plan, 2008–2011*. 4 Apr. 2008. PDF file.

———. "Facts and Statistics." Web. 17 Sept. 2013.

———. "Oil Sands Projects and Upgraders." Web. 3 Sept. 2013.

Almack, Kaitlin. "Thoughts of an Anthropologist." *The Enpipe Line: 70,000 km of Poetry Written in Resistance to the Northern Gateway Pipelines Proposal*. Smithers, BC: Creekstone P, 2012. Web.

Althusser, Louis. "Ideology and Ideological State Apparatuses (Notes towards an Investigation)." *Lenin and Philosophy*. Trans. Ben Brewster. New York: Monthly Review P, 1971. Print.

Anderson, Benedict. *Imagined Communities: Reflections on the Origin and Spread of Nationalism*. 1991. New York: Verso, 2000. Print.

Angenot, Marc. "What Can Literature Do? From Literary Sociocriticism to a Critique of Social Discourse." *Yale Journal of Criticism: Interpretation in the Humanities* 17.2 (2004): 217–31. Print.

Angus, Ian. *A Border Within: National Identity, Cultural Plurality, and Wilderness*. Montreal and Kingston: McGill-Queen's UP, 1997. Print.

———. "Continuing Dispossession: Clearances as a Literary and Philosophical Theme." www.academia.edu. Web. 24 July 2015.

———. *(Dis)Figurations: Discourse/Critique/Ethics.* New York: Verso, 2000. Print.

———. *Emergent Publics: An Essay on Social Movements and Democracy.* Winnipeg: Arbeiter Ring Publishers. 2001. Print.

Aristotle. *The Art of Rhetoric.* Trans. Hugh Lawson-Tancred. Toronto: Penguin, 1991. Print.

Armstrong, Jeanette. "Literature of the Land: An Ethos for These Times." Association for Commonwealth Literature and Language Studies Fourteenth Triennial Conference. University of British Columbia. 18 Aug. 2007. Lecture.

Arsenault, Chris. "CANADA: Govt Threatens Tar Sands Activists with Anti-Terror Laws." *IPS: Inter Press Service News Agency.* 20 Oct. 2009. Web.

ATB Financial. Advertisement. CBC. 27 Aug. 2007. Television.

Atkinson, Ted. "Blood Petroleum: *True Blood,* the BP Oil Spill, and Fictions of Energy/Culture." *Journal of American Studies* 47.1 (2013): 213–29. Print.

Auerbach, Nina. *Our Vampires, Ourselves.* Chicago: U of Chicago P, 1995. Print.

Ayotte, Devin. "From the Depths: Social Commentary and the Emerging 'Rig Pig' Aesthetic." Petrocultures: Oil, Energy, and Culture. Edmonton, AB. 7 Sept. 2013. Conference Presentation.

Babiak, Todd. "Consent, Not Dissent, the Order of the Day." *Edmonton Journal* 7 Feb. 2008: C1. Print.

Babiuk, Colin. *Oil Sands and the Earth: Framing the Environmental Message in the Print News Media.* Charleston, SC: VDM Verlag Dr. Muller, 2008. Print.

Ball, David P. "First Nation Challenges Shell Tar Sands Expansion." *Indian Country Today* 2 Oct. 2012. Web.

Barney, Darin. "'Who We Are and What We Do': Canada as a Pipeline Nation." Abstract for MLA Presentation. https://commons.mla.org, 6 Jan. 2015. Web.

Bavington, Dean. *Managed Annihilation: An Unnatural History of the Newfoundland Cod Collapse.* Vancouver: UBC P, 2010. Print.

Bayne, Erin M., and Brenda C. Dale. "Effects of Energy Development on Songbirds." *Energy Development and Wildlife Conservation in Western North America.* Ed. David E. Naugle. Washington, DC: Island P, 2011. 95–114. Print.

Beamish, Michael. *Simple Simon.* interPLAY Visual and Performing Arts Festival: Aug. 8–10, 2008. Fort McMurray, AB. 10 Aug. 2008. Performance.

Bennett, Dean. "Province Defends Keystone XL in $30,000 *New York Times* Ad." *Edmonton Journal* 18 Mar. 2013: A7. Print.

Bennett, Jane. *Vibrant Matter: A Political Ecology of Things.* Durham, NC: Duke UP, 2010. Print.

Berlant, Lauren. "Slow Death (Sovereignty, Obesity, and Lateral Agency). *Critical Inquiry* 33.4 (2007): 754–80. Print.

Bertell, Susan. "The Standard is Zero." *Alberta Views* Nov. 2007: 18–19. Print.

Blackwell, Richard. "Alberta Cancels Funding For Carbon Capture Project." *Globe and Mail* 25 Feb. 2013. Web.

Blanchot, Maurice. *The Unavowable Community*. 1983. Trans. Pierre Joris. New York: Station Hill P, 1988. Print.

Boswell, Randy, and Michelle Butterfield. "Oilsands Under Fire in U.K.: Investment Fund Puts Heat on BP, Shell to Scale Back Plans on 'Risky' Resource." *Edmonton Journal* 15 Sept. 2008: A1. Print.

Bott, Robert. *Canada's Oil Sands*. 2nd ed. Ed. David M. Carson and Tami Hutchinson. Sept. 2007. Canadian Centre for Energy Information. Web.

Bouchard, Gérard. "What is Interculturalism?" *McGill Law Journal–Revue de droit de McGill*. 56.2 (2011): 435–68. Print.

Bowers, C.A. *Educating for Eco-Justice and Community*. Athens: U of Georgia P, 2001. Print.

———. *Mindful Conservatism*. Boulder, CO: Rowman and Littlefield Publishers, 2003. Print.

Braz, Albert. "Wither the White Man: Charles Mair's 'Lament for the Bison.'" *Canadian Poetry* 49 (Fall/Winter 2001). Web.

Brooymans, Hanneke. "Fort Chipewyan Wants Answers about Cancer Rates." *Edmonton Journal* 5 Sept. 2010: A1. Print.

———. "Oilsands Boosts Toxic Metals in Athabasca Watershed: Study." *Edmonton Journal* 31 Aug. 2010: A3. Print.

———. "Oilsands' Newest Project: A Greener Image." *Edmonton Journal* 14 Oct. 2007: A1. Print.

Brydon, Diana. "Canada and Postcolonialism: Questions, Inventories, Futures." *Is Canada Postcolonial? Unsettling Canadian Literature*. Ed. Laura Moss. Waterloo, ON: Wilfrid Laurier UP, 2003. 49–77. Print.

———. "Metamorphoses of a Discipline: Rethinking Canadian Literature within Institutional Contexts." *Trans.Can.Lit: Resituating the Study of Canadian Literature*. Ed. Smaro Kamboureli and Roy Miki. Waterloo, ON: Wilfrid Laurier UP, 2007. 1–16. Print.

Burke, Kenneth. *A Rhetoric of Motives*. Berkeley: U of California P, 1969. Print.

Burkett, Maxine. "Climate-Induced Migration and the Challenge of Statehood: Defining the Problem, Identifying Solutions." Changing Nature: Migrations, Energies, Limits. Association for the Study of Literature and the Environment (ASLE) Tenth Biennial Conference. University of Kansas, Lawrence. 30 May 2013. Plenary Address.

Canada. *Treaty No. 8, Made June 21, 1899, and Adhesions, Reports, etc.* Ottawa: Roger Duhamel, Queen's Printer and Controller of Stationary, 1966. www.treaty8.ca. Web.

Canadian Association of Petroleum Producers (CAPP). "Upstream Dialogue: The Facts on Oilsands." June 2012. Pamphlet. Print.

Canadian Press. "Alberta Energy Regulator Shutting off Discussions, Critics Say." CBC *News*, 18 May 2014. Web.

———. "Caribou 'Lifeboat' Recovery Pen Considered for Oilsands Region." CBC *News*, 10 Oct. 2012. Web.

———. "Government Hones Oilsands Message with Focus Groups." CBC *News*, 18 Feb. 2013. Web.

———. "Oil and Gas Ad Campaign Cost Feds $40M at Home and Abroad." CBC *News*, 27 Nov. 2013. Web.

———. "Ottawa to Consider Penning Oilsands Caribou behind Fence." *The Tyee* 10 Oct. 2012. Web.

Caputo, John D. *Radical Hermeneutics: Repetition, Deconstruction, and the Hermeneutic Project.* Bloomington: Indiana UP, 1987. Print.

Cariou, Warren. "An Athabasca Story." *Lake: A Journal of Arts and Environment* (Spring 2012): 70–75. Print.

———. "Dances with Rigoureau." *Troubling Tricksters: Revisioning Critical Conversations.* Ed. Deanna Reder and Linda Morra. Waterloo, ON: Wilfrid Laurier UP, 2010. 157–67. EBSCO. Web.

———."Tarhands: A Messy Manifesto." *Imaginations: Journal of Cross-Cultural Image Studies* 3.2 (2012). Web.

Cattaneo, Claudia. "'A Battle for the Country's Future': How Joe Oliver Has Become Canada's Energy Pitchman." *Financial Post* 19 Apr. 2013. Web.

———. "Northern Gateway's Biggest Risk to Canada Is Not Approving Pipeline: Enbridge." *Financial Post* 17 June 2013. Web.

CBC News. "Healing Walk Draws Hundreds to Protest Oil Sands." *cbc.ca*, 4 Aug. 2012. Web.

———. "Joe Oliver Slams Scientist's Oilsands Claims As 'Nonsense.'" *cbc.ca*, 24 Apr. 2013. Web.

CEMA. "Backgrounder: CEMA Delivers Oilsands Mine End Pit Lake Guidance Document." 4 Oct. 2012. Web.

Chan, Kai M.A., Anne Salomon, Eric B. Taylor, et al. "Open Letter on the Joint Review Panel Report Regarding the Northern Gateway Project." 26 May 2014. Web.

Chapman, Ken. "The Next North Needs New Ways of Thinking." *Oil Sands Ken*, 30 July 2013. Web.

———. Presentation to the Petrocultures Conference. Fort McMurray, AB. 10 Sept. 2012. Question and Answer.

Chapman, K.J., et al. "The Oil Sands Survey: Albertans' Values Regarding Oil Sands Development." Cambridge Strategies Inc., Edmonton, AB, 2008. Print.

Chastko, Paul. *Developing Alberta's Oil Sands: from Karl Clark to Kyoto.* Calgary: U of Calgary P, 2004. Print.

Cheadle, Bruce. "CRA Charity Audits: Study Says Rules for Political Advocacy Are Unclear, Unfair." *Huffington Post* 25 Mar. 2015. Web.

Chisholm, Dianne. "The Art of Ecological Thinking." *Interdisciplinary Studies in Literature and the Environment* 18.3 (2011): 569–93. Print.

Christian, Carol. "Killed Workers Remembered." *Fort McMurray Today* 29 Apr. 2009. Web.

Chung, Emily. "Humpback Whale Losing 'Threatened' Status amid Northern Gateway Concerns." CBC *News*, 22 Apr. 2014. Web.

Clark, Karl. *The Athabasca Tar Sands*. May 1949. University of Alberta Archives 76-93-16. Print.

———. *Commercial Development Feasible for Alberta's Bituminous Sands*. Research Council of Alberta. Contribution 26, 1929. University of Alberta Archives 76-93-20. Print.

———. Letters. University of Alberta Archives 84–43. Print.

Clark, Timothy. *The Cambridge Introduction to Literature and the Environment*. New York: Cambridge UP, 2011. Print.

Code, Lorraine. *Ecological Thinking: The Politics of Epistemic Location*. Toronto: Oxford UP, 2006. Print.

Coleman, Daniel. *In Bed with the Word: Reading, Spirituality, and Cultural Politics*. Edmonton: U of Alberta P, 2009. Print.

Collyer, Dave. "Oil, Gas Industry Is Making Environmental Progress." *Edmonton Journal* 11 Aug. 2010: A15. Print.

Comaroff, Jean, and John Comaroff. "Alien-Nation: Zombies, Immigrants, and Millennial Capitalism." *South Atlantic Quarterly* 101.4 (2002): 779–805. Print.

Corlett, William. *Community without Unity: A Politics of Derridian Extravagance*. 1989. Durham, NC: Duke UP, 1993.

Cotter, John. "Sherritt Delays Coal Gasification Plant." *Canadian Press*, 30 Aug. 2008. Web.

Dannenberg, Hilary P. *Coincidence and Counterfactuality: Plotting Time and Space in Narrative Fiction*. Lincoln: U of Nebraska P, 2008. Web.

Das, Satya. *Green Oil*. Edmonton: Sextant Publishing, 2009. Print.

Davidsen, Conny. "Looking Back on Canadian Energy, Oil/Tar Sands, and Climate Change Discourse(s)." Under Western Skies 2: Environment, Community, and Culture in North America. Mount Royal University, Calgary, AB. 13 Oct. 2012. Conference Presentation.

Davis, Arthur. "Did George Grant Change His Politics?" *Athens and Jerusalem: George Grant's Theology, Philosophy, and Politics*. Ed. Ian Angus, Ron Dart, and Randy Peg Peters. Toronto: U of Toronto P, 2006. 62–79. Print.

Dead Ducks. Dir. Brenda Longfellow. Online film. *Vimeo*, 3 Mar. 2012. Web.

De Certeau, Michel. *The Practice of Everyday Life*. 1984. Trans. Steve Rendall. Berkeley: U of California P, 1988. Print.

Dembecki, Geoff. "Harper Budget Has $8 Million to Restrict 'Political Activities' of Charities." *The Tyee* 29 Mar. 2012. Web.

Den Tandt, Michael. "Tutu's Oilsands Sermon Built on Emotion." *Edmonton Journal* 2 June 2014: A11. Print.

Derrida, Jacques. *The Gift of Death*. 1992. Trans. David Wills. Chicago: U of Chicago P, 1996. Print.

De Souza, Mike. "Caribou at Risk in Oilsands Regions." *Edmonton Journal* 7 July 2012: A13. Print.

Dickinson, Adam. "Lyric Ethics: Ecocriticism, Material Metaphoricity, and the Poetics of Don McKay and Jan Zwicky." *Canadian Poetry* 55 (Fall/Winter 2004): 34–52. Web.

Dillingham, William B. *Rudyard Kipling: Hell and Heroism.* Gordonsville, VA: Palgrave Macmillan, 2005. Web.

Dorow, Sara, and Goze Dogu. "The Spatial Distribution of Hope in and beyond Fort McMurray." *Ecologies of Affect: Placing Nostalgia, Desire, and Hope.* Waterloo, ON: Wilfrid Laurier UP, 2011. Print.

Dorow, Sara, and Naomi Krogman. "Social Impacts of Oil Sands Development: Research on Fort McMurray." School of Energy and the Environment, University of Alberta. 12 Feb. 2010. Lecture.

Dorow, Sara, and Sara O'Shaugnessy. "Fort McMurray, Wood Buffalo, and the Oil/Tar Sands: Revisiting the Sociology of 'Community': Introduction to the Special Issue." *CJS: Canadian Journal of Sociology/Cahiers Canadiens de Sociologie* 38.2 (2013): 121–40. Print.

Dryzek, John S. *The Politics of the Earth: Environmental Discourses.* 1997. Oxford: Oxford UP, 2005. Print.

Dubois, Frédéric, Marc Tessier, and David Widgington, eds. *Extraction! Comix Reportage.* Montreal: Cumulus P, 2007. Print.

Edwards, James S. Foreword. *Athabasca Oil Sands: From Laboratory to Production: The Letters of Karl A. Clark, 1950–66.* Ed. Mary Clark Sheppard. Sherwood Park, AB: Geoscience Publishing, 2005. Print.

Ells, S.C. *Recollections of the Development of the Athabasca Oil Sands.* Ottawa: Department of Mines and Technical Surveys, 1962. Library and Archives Canada, AMICUS 20865726. Print.

Ermine, Willie. "The Ethical Space of Engagement." *Indigenous Law Journal* 6.1 (2007): 193–203. Print.

Ethical Oil: A Choice We Have to Make. www.ethicaloil.org. Web. 1 Aug. 2011.

Evans, C.D., and L.M. Shyba. *5000 Dead Ducks: A Novel about Lust and Revolution in the Oilsands.* Calgary: Durance Vile, 2011. Print.

Fehr, Joy Anne. "(Re)Writing Alberta." Diss. University of Calgary, 2005. Print.

Fey, Tina. *Bossypants.* New York: Little, Brown and Company, 2011. Print.

Finch, David. *Pumped: Everyone's Guide to the Oilpatch.* Calgary: Fifth House, 2007. Print.

Findlay, Len. "Is Canada a Postcolonial Country?" *Is Canada Postcolonial? Unsettling Canadian Literature.* Ed. Laura Moss. Waterloo, ON: Wilfrid Laurier UP, 2003. 297–99. Print.

Fluker, Shaun. "*R v Syncrude Canada*: A Clash of Bitumen and Birds." *Alberta Law Review* 49.1 (2012): 237–44. Print.

———. "Wilderness as the Antithesis of Law: Implications for Environmental Protection." Land Use and Wilderness Panel, Under Western Skies 2: Environment, Community, and Culture in North America. Mount Royal University, Calgary, AB. 11 Oct. 2012. Conference Presentation.

"Foreign Influence Foolish Target." Editorial. *Edmonton Journal* 10 Jan. 2012: A12. Print.

Foucault, Michel. *Fearless Speech*. Ed. Joseph Pearson. Los Angeles: Semiotext(e), 2001. Print.

Frankfurt, Harry. "On Bullshit." Princeton University. Web. 10 Aug. 2010.

Fuller, Steve. "Just Bullshit." *Bullshit and Philosophy*. Ed. Gary L. Hardcastle and George A. Reisch. Chicago: Open Court, 2006. 241–58. Print.

Gailus, Jeff. "Black Swans: Think Oil Sands Companies Are Certain to Clean Up after Themselves? Think Again. How Citizens Could Be Stuck with the $15-Billion Bill." *Alberta Views* Oct. 2012: 34–39. Print.

———. *Little Black Lies: Corporate and Political Spin in the Global War for Oil*. Calgary: Rocky Mountain Books, 2011. Print.

Gauthier, Alexandre. "L'illusion bitumineuse." *Canadian Literature* (Winter 2010). Web.

Glanz, James. "Google Details, and Defends, Its Use of Electricity." *New York Times* 9 Sept. 2011. Web.

Gleicher, Faith, et al. "With an Eye Toward the Future: The Impact of Counterfactual Thinking on Affect, Attitudes, and Behavior." *What Might Have Been: The Social Psychology of Counterfactual Thinking*. Ed. Neal J. Roese and James M. Olsen. Mahwah, NJ: Lawrence Erlbaum Associates, 1995. 283–304. Print.

Gismondi, Mike, and Debra J. Davidson. *Challenging Legitimacy at the Precipice of Energy Calamity*. New York: Springer, 2011. Web.

Goldie, Terry. *Fear and Temptation: The Image of the Indigene in Canadian, Australian, and New Zealand Literatures*. Montreal and Kingston: McGill-Queen's UP, 1989.

Gordon, Jon. "Displacing Oil: Towards 'Lyric' Re-presentations of the Alberta Oil Sands." *Countering Displacements: The Creativity and Resilience of Indigenous and Refugee-ed Peoples*. Ed. Daniel Coleman, Erin Goheen Glanville, Wafaa Hasan, and Agnes Kramer-Hamstra. Edmonton: U of Alberta P, 2012. 1–30. Print.

———. "Rethinking Bitumen: From 'Bullshit' to a 'Matter of Concern.'" *Imaginations: Journal of Cross-Cultural Image Studies* 3.2 (2012). Web.

Gorrell, Donna. "The Rhetorical Situation Again: Linked Components in a Venn Diagram." *Philosophy and Rhetoric* 30.4 (1997): 395–412. Web.

Gosselin, Pierre, et al. *The Royal Society of Canada Expert Panel: Environmental and Health Impacts of Canada's Oil Sands Industry*. Ottawa: Royal Society of Canada, 2010. Web.

Government of Alberta. Environment and Parks. *Oil Sands Mining Development and Reclamation*. www.esrd.alberta.ca. Web. 11 June 2015.

———. *Oil Sands Information Portal*. http://oilsands.alberta.ca. Web. 5 July 2013.

———. *The Oil Sands Sustainable Development Secretariat*. www.energy.alberta.ca. Web. 25 Aug. 2008.

———. Times Square Advertisement. *Calgary Sun*. Web. 9 Sept. 2010.

Grant, George. *English-Speaking Justice*. 1974. Toronto: Anansi, 1998.

———. *Lament for a Nation: The Defeat of Canadian Nationalism*. 1965; repr. Montreal and Kingston: McGill-Queen's UP, 2005.

———. *Philosophy in the Mass Age*. 1959. Ed. William Christian. Toronto: U of Toronto P, 1995.

———. *Technology and Empire*. Toronto: Anansi, 1969.

———. *Technology and Justice*. Toronto: Anansi, 1986. Print.

Grant, Jennifer, Simon Dyer, and Dan Woynillowicz. "Fact or Fiction: Oil Sands Reclamation." www.pembina.org, May 2008. Web.

Grant, Sheila. Afterword. *Lament for a Nation: The Defeat of Canadian Nationalism*. 1997. Montreal and Kingston: McGill-Queen's UP, 2005. Print.

Grant-Davie, Keith. "Rhetorical Situations and Their Constituents." *Rhetoric Review* 15.2 (1997): 264–79. Web.

Graveland, Bill. "Stick to Keystone Facts, US Governor Says." *Edmonton Journal* 29 Mar. 2013: A10. Print.

Grubisic, Katia. "Savage Nations Roam o'er Native Wilds": Charles Mair and the Ecological Indian." *Studies in Canadian Literature* 30.1 (2005): 58–83. Web.

Hatch, Christopher, and Matt Price. "Canada's Toxic Tar Sands: The Most Destructive Project on Earth." Environmental Defence. Feb. 2008. Web.

Hebblewhite, Mark. "Effects of Energy Development on Ungulates." *Energy Development and Wildlife Conservation in Western North America*. Ed. David E. Naugle. Washington, DC: Island P, 2011. 71–94. Print.

Heidegger, Martin. *The Question Concerning Technology and Other Essays*. Trans. William Lovitt. San Francisco: Harper Colophon, 1977. Print.

Henderson, Matthew. *The Lease*. Toronto: Coach House Books, 2012.

Henton, Darcy. "If Syncrude Convicted, Oilsands 'Doomed': Defence." *Edmonton Journal* 29 Apr. 2010: A1. Print.

Hobbes, Thomas. *Leviathan*. 1651. Ed. C.B. Machperson. London: Penguin, 1985. Print.

The Holy Bible: New Revised Standard Version. Grand Rapids, MI: Zondervan Bible Publishers, 1990. Print.

Hubbard, Tasha. "'The Buffaloes are Gone' or 'Return: Buffalo'?: The Relationship of the Buffalo to Indigenous Creative Expression." *Canadian Journal of Native Studies* 29.1&2 (2009): 65–85. Web.

Huggan, Graham, and Helen Tiffin. *Postcolonial Ecocriticism: Literature, Animals, Environment*. New York: Routledge, 2010. Print.

Humberman, Irwin. *The Place We Call Home: A History of Fort McMurray as Its People Remember, 1778–1980*. Edmonton: Historical Book Society of Fort McMurray, 2001. Print.

Hunter, William H., et al. *Our Fair Share, Report of the Alberta Royalty Review Panel*. 18 Sept. 2007. Print.

Innes, Robert Alexander. *Elder Brother and the Law of the People: Contemporary Kinship and Cowessess First Nation*. Winnipeg: U of Manitoba P, 2013. Print.

Iribarne, J. "Cetology." *Poems for an Oil-Free Coast*. Victoria, BC: Red Tower Bookworks, 2012. Print.

James, Peter. "Oil Spills Can Benefit Economy, Panel Told." *Prince George Citizen* 27 Feb. 2013. Web.

Jameson, Fredric. *The Seeds of Time*. New York: Columbia UP, 1996. Print.

Jamieson, JoAnn P., John M. Olynyk, and Trevor C. Ference. "Alberta's New Energy Regulator: What Does it Mean for Project Development?" *Energy Law Bulletin* 16 Nov. 2012. Web.

Katinas, Tom. "An Apology from Syncrude—and a Promise to Do Better." www.syncrude.ca, 2 May 2008. Web.

Kavanagh, James H. "Ideology." *Critical Terms for Literary Study*. 2nd ed. Ed. Frank Lentricchia and Thomas McLaughlin. Chicago: U of Chicago P, 1995. Print.

Kaye, Frances W. *Good Lands: A Meditation and History on the Great Plains*. Edmonton: Athabasca UP, 2011. Print.

Keller, James. "Ottawa Backs Gateway despite US Salvo at Enbridge." *Edmonton Journal* 19 July 2012: A2. Print.

Kelly, Erin N., David W. Schindler, et al. "Oil Sands Development Contributes Polycyclic Aromatic Compounds to the Athabasca River and Its Tributaries." *Proceedings of the National Academy of Sciences USA* 106 (2009): 22346–22351. Print.

Kennedy, Jessica, Martin Ignasiak, and Sander Duncanson. "Alberta First Nations Challenge Constitutionality of Federal Regulatory Reforms." *Osler*, 9 Jan. 2013. Web.

Kertzer, Jonathan. "Bio-Critical Essay." *The Rudy Wiebe Papers, First Accession: An Inventory of the Archive at the University of Calgary Libraries*. Calgary: U of Calgary P, 1986. Print.

Kipling, Rudyard. "If." www.swarthmore.edu. Web. 16 June 2009.

Klaszus, Jeremy. "Athabasca Blues." *Alberta Views* Nov. 2007: 30–35. Print.

Kleiss, Karen. "'Tough' Talk over Red Ink." *Edmonton Journal* 20 Dec. 2012: A1. Print.

Klinkenberg, Marty. "Health Care Concerns Force Fort McMurray Couple to Abandon City They Love." *Edmonton Journal* 4 Sept. 2013. Web.

———. "Rethinking the Oilsands Impact." *Edmonton Journal* 26 Jan. 2013: A7. Print.

King, Thomas. "Alberta Oil Sands Land." www.ndp.ca. Web. 12 June 2008.

———. *All My Relations: An Anthology of Contemporary Canadian Native Fiction*. Toronto: McClelland & Stewart, 1990. Print.

———. *The Truth About Stories: A Native Narrative*. Toronto: Anansi, 2003. Print.

Kincaid, Jamaica. *A Small Place*. New York: Penguin, 1988. Print.

Kopecky, Arno. "Game Changer." *Alberta Views* June 2014: 34–41. Print.

Kranz, L.A. Review of *Godless But Loyal to Heaven*. *goodreads.com*, 1 Feb. 2013. Web.

Kroetsch, Robert. *Alberta*. 1967. Edmonton: NeWest P, 1993. Print.

———. "Is This a Real Story or Did You Make It Up?" *Eighteen Bridges: Stories that Connect* Fall 2010: 18–19. Print.

Lamontague, Patrick. "Political Cartoon." *Edmonton Journal* 14 Sept. 2013: A22. Print.

Lamphier, Gary. "Notley's NDP Stakes Middle Ground." *Edmonton Journal* 27 June 2015: C1. Print.

———. "Time to Act on Oilsands Emissions." *Edmonton Journal* 18 June 2015: B1, B3. Print.

Land of Oil and Water. Dir. Neil McArthur and Warren Cariou. 2009. Film.

Latour, Bruno. "The Politics of Explanation: An Alternative." *Knowledge and Reflexivity: New Frontiers in the Sociology of Knowledge*. Ed. Steve Woolgar. Beverly Hills, CA: Sage Publications, 1988. Print.

———. *What Is the Style of Matters of Concern?* 2005 Spinoza Lectures. Department of Philosophy University of Amsterdam. VanGorcum, 2008. Print.

———."Why Has Critique Run Out of Steam? From Matters of Fact to Matters of Concern." *Critical Inquiry* 30.2 (2004): 225–48. Print.

Lavis, Maria. "Lingerers On." *The Enpipe Line: 70,000 km of Poetry Written in Resistance to the Northern Gateway Pipelines Proposal*. Smithers, BC: Creekstone P, 2012. Web.

Leach, Andrew. "Don't Let EU Define Our Emissions Policy." *Edmonton Journal* 2 Mar. 2011: A15. Print.

———. "A False Hope Underpins the Mine Financial Security Program." *Alberta Oil: The Business of Energy*, 1 June 2011. Web.

Leclerc, Christine, ed. *The Enpipe Line: 70,000 km of Poetry Written in Resistance to the Northern Gateway Pipelines Proposal*. Smithers, BC: Creekstone P, 2012. Web.

Lefsrud, Lianne. "Who Killed the Golden Goose? Stakeholders' Meaning Making for Alberta's Oil Sands." Petrocultures: Oil, Energy, and Culture. Edmonton, AB. 8 Sept. 2013. Conference Presentation.

LeMenager, Stephanie. *Living Oil: Petroleum Culture in the American Century*. New York: Oxford UP, 2014. Print.

Lemphers, Nathan. "Open for Business, Closed to Criticism." *The Mark: The People and Ideas Behind the Headlines*, 11 Jan. 2012. Web.

Leonard, David W. "Charles Mair and the Settlement of 1899." *Through the Mackenzie Basin: An Account of the Signing of Treaty No. 8 and the Scrip Commission, 1899*. 1908. Edmonton: U of Alberta P and the Edmonton District Historical Society, 1999. xv–xxxvii. Print.

Levant, Ezra. *Ethical Oil: The Case for Canada's Oil Sands*. Toronto: McClelland & Stewart, 2010. Print.

———. "Shed No Tears for Ft. McMurray's Ducks." *National Post* 5 May 2008. Web.

Levant, Ezra, and Andrew Nikiforuk. "Debate" Q. CBC Radio. Web.

Lewis, Jeff. "How Suncor Climbed atop the List of Canada's Biggest Companies." *Financial Post Magazine* 18 June 2013. Web.

Lillebuen, Steve. "Native Groups Sue Gov't over Oilsands." *Edmonton Journal* 5 June 2008: A3. Print.

Mahaffy, Cheryl. "The Hidden Face of Prosperity." *Alberta Views* Nov. 2007: 36–41. Print.

Maher, Stephen. "We're Telling Our Customer He's Wrong." *Edmonton Journal* 28 Sept. 2013: A18. Print.

Mair, Charles. "The Last Bison." *Tecumseh, A Drama and Canadian Poems. Dreamland and Other Poems. The American Bison. Through the Mackenzie Basin. Memoirs and Reminiscences*. Toronto: The Radisson Society, 1926. 237–42. Print.

———. *Through the Mackenzie Basin: An Account of the Signing of Treaty No. 8 and the Scrip Commission, 1899.* 1908. Edmonton: U of Alberta P and the Edmonton District Historical Society, 1999. Print.

Major, Alice. "The Politics of Art." *Alberta Views* Mar. 2013: 18. Print.

Manufactured Landscapes. Dir. Jennifer Baichwal. Zeitgeist Films Ltd., 1996. DVD.

Marsden, William. *Stupid to the Last Drop.* Toronto: Knopf, 2007. Print.

Mason, Travis. "Listening at the Edge: Homage and Ohmage in Don McKay and Ken Babstock." *Studies in Canadian Literature* 33.1 (2008): 77–96. Web.

Mayer, Ariel. "The 'Advocacy Chill': CRA Audits and Political Activity." *The Public Policy & Governance Review* 24 Nov. 2014. Web.

McDermott, Vincent. "CEMA Receives Funding for One Year." *Fort McMurray Today* 28 Jan. 2014. Web.

McLean, Archie, and Hanneke Brooymans. "Premier Wants Truth about Toxins." *Edmonton Journal* 2 Sept. 2010: A1. Print.

Meisner, Natalie. *Boom, Baby.* Staged Reading at Under Western Skies 3. Mount Royal University, Calgary, AB. 10 Sept. 2014. Work in Progress.

Mitrovica, Andrew. "Rex Murphy, the Oilsands and the Cone of Silence." *iPolitics,* 10 Feb. 2014. Web.

Mockler, Kathryn. "Pipeline." *The Enpipe Line: 70,000 km of Poetry Written in Resistance to the Northern Gateway Pipelines Proposal.* Smithers, BC: Creekstone P, 2012. Web.

Mookerjea, Sourayan. "Epilogue: Through the Utopian Forest of Time." *CJS: Canadian Journal of Sociology/Cahiers Canadiens de Sociologie* 38.2 (2013): 233–54. Print.

Moore, Dr. Patrick. "Attacking Oilsands a 'Dirty' Trick: Canada Far ahead of Other Oil-Exporting Nations on Social and Environmental Fronts." Op-ed. *Edmonton Journal* 8 Dec. 2011: A19. Print.

Munro, Margaret. "Academics Seek Freeze on Oilsands Projects." *Edmonton Journal* 26 June 2014: A9. Print.

Murphy, Rex. "Humane Environmentalism." *Alberta Views* July/Aug. 2012: 20. Excerpted from "A Triumph for the Human Environment." *National Post* 17 Mar. 2012. Print.

Murphy, Verna. "Berating the Home that Feeds." *Alberta Views* May 2014: 24. Reprinted from "Fellow Cape Bretoner's Rant about Fort McMurray Goes Viral." *Cape Breton Post* 14 May 2013. Print.

Nancy, Jean-Luc. *Being Singular Plural.* Trans. Robert D. Richardson and Anne E. O'Byrne. Stanford: Stanford UP, 2000. Print.

———. *The Inoperative Community.* Ed. Peter Connor. Trans. Peter Connor et al. Minneapolis: U of Minnesota P, 1991. Print.

———. "The Unsacrificeable." Trans. Richard Livingstone. *Yale French Studies* 79 (1991): 20–38. Print.

National Arts Centre/Centre national des Arts. Review of *Fort Mac* by Marc Prescott. *National Arts Centre/Centre national des Arts,* 28 Aug. 2007. Web.

National Energy Board (NEB). *Connections: Report of the Joint Review Panel for the Enbridge Northern Gateway Project: Volume 1*. 2013. Web.

Nicholls, Liz. "Oil Boom Takes to the Stage." *Edmonton Journal* 12 Apr. 2007. Web.

Nikiforuk, Andrew. *The Energy of Slaves: Oil and the New Servitude*. Vancouver: Greystone Books, 2012. Print.

———. *Tar Sands: Dirty Oil and the Future of a Continent*. Vancouver: Greystone Books, 2008. Print.

Nixon, Rob. *Slow Violence and the Environmentalism of the Poor*. Cambridge, MA: Harvard UP, 2011. Print.

Northern Gateway. *Northern Gateway*. www.gatewayfacts.ca. Web. 15 Aug. 2014.

Northern Gateway Pipeline. Advertisement. *The Walrus* Dec. 2013: 3–4. Print.

Nuttall, Jeremy J. "Pattern of Charity Audit Bias 'Disturbing' Says Free Speech Advocate." *The Tyee* 22 Oct. 2014. Web.

Oliver, Joe. "An Open Letter from Natural Resources Minister Joe Oliver." *Globe and Mail* 9 Jan. 2012. Web.

O'Neil, Peter. "Feds Slam 'Radical' Greens: 'Jet-Setting Celebrities' Trying to Hijack Regulatory System, Resource Minister Says." *Edmonton Journal* 9 Jan. 2012: A1. Print.

———. "Teachers Accused of Pipeline 'Propaganda.'" *Edmonton Journal* 4 Oct. 2012: A7. Print.

O'Shaugnessy, Sara, and Naomi T. Krogman. "Gender as Contradiction: From Dichotomies to Diversity in Natural Resource Extraction." *Journal of Rural Studies* 27.2 (2011): 134–43. Print.

Outka, Paul. "Roundtable: Building the Environmental Humanities." Changing Nature: Migrations, Energies, Limits. Association for the Study of Literature and the Environment Tenth Biennial Conference. University of Kansas, Lawrence. 31 May 2013. Conference Presentation.

Paris, Max. "Inside Politics Blog." *cbc.ca*, 27 Sept. 2011. Web.

Porter, James. "Intertextuality and the Discourse Community." *Rhetoric Review* 5.1 (1986): 34–47. Web.

Pratt, Larry. *The Tar Sands: Syncrude and the Politics of Oil*. Edmonton: Hurtig, 1976. Print.

Pratt, Sheila. "Air Cannons Not Scaring Birds Away from Tailings Ponds: Noise No Deterrent, Monitoring Program Finds." *Edmonton Journal* 12 July 2013. Web.

———. "Appeal Court Backs Cree Nation." *Edmonton Journal* 5 June 2013: A6. Print.

———. "'Eyes of the World' on Oilsands Bird Safety." *Edmonton Journal*. 30 May 2014: A4. Print.

———. "Funding Cut in Half for Oilsands Environmental Group." *Edmonton Journal* 13 Dec. 2012. Web.

———. "Policy Seeks to Minimize Destruction of Wetlands." *Edmonton Journal* 11 Sept. 2013: A4. Print.

Prescott, Marc. *Fort Mac*. Saint-Boniface, MB: Rouge, 2009. Print.

"Progress." OED *Online*. Oxford UP, June 2015. Web.

Quist, Jennifer. "Back." *North Word: A Literary Journal of Canada's North* 2.2 (2012): 4–6. Print.

Rainey, Lawrence. *Institutions of Modernism: Literary Elites and Public Culture.* New Haven, CT: Yale UP, 1998. Print.

Red Tower Bookworks. Review Copy Descriptive Insert. *Poems for an Oil-Free Coast.* Victoria, BC: Red Tower Bookworks, 2012. Print.

Reid, Ian. "Public Library Data Service 2012 Statistical Report: Characteristics and Trends." *Public Libraries Online,* 18 Dec. 2012. Web.

Rethink Alberta. "Rethink Alberta." Online video. *YouTube,* 15 July 2010. Web.

Richardson, Lee. *The Oil Sands: Toward Sustainable Development.* Report of the Standing Committee on Natural Resources. Mar. 2007. Web.

Rocan, Susan. "Interview with Richard Van Camp." *mywithershins,* 11 Nov. 2012. Web.

Rowe, Stan. "'Homo Ecologicus'—A Wiser Name." *Earth Alive: Essays on Ecology.* Edmonton: NeWest P, 2006. 116–19. Print.

Rowley, Mari-Lou. "In the Tar Sands, Going Down." *Regreen: New Canadian Ecological Poetry.* Ed. Madhur Anand and Adam Dickinson. Sudbury, ON: Your Scrivener P, 2009. 63–66. Print.

Ryan, Terre. "Dead Worker Heroes and Disastrous Energy Discourse." Changing Nature: Migrations, Energies, Limits. Association for the Study of Literature and the Environment Tenth Biennial Conference. University of Kansas, Lawrence. 30 May 2013. Conference Presentation.

Sandilands, Catriona. *The Good-Natured Feminist: Ecofeminism and the Quest for Democracy.* Minneapolis: U of Minnesota P, 1999. Print.

Sawatsky, Melissa. "Say." *The Enpipe Line: 70,000 km of Poetry Written in Resistance to the Northern Gateway Pipelines Proposal.* Smithers, BC: Creekstone P, 2012. Web.

Schindler, David. "Climate Unchanged." *Alberta Views* Sept. 2013: 19. Print.

"Science." OED *Online.* Oxford UP, June 2015. Web.

Severson-Baker, Chris, Jennifer Grant, and Simon Dyer. "Taking the Wheel: Correcting the Course of Cumulative Environmental Management in the Athabasca Oil Sands." Pembina Institute. 18 Aug. 2008. PDF file.

Sheppard, Mary Clark, ed. *Oil Sands Scientist: The Letters of Karl A. Clark 1920–1949.* Edmonton: U of Alberta P, 1989. Print.

Shiell, Leslie, and Colin Busby. "Greater Saving Required: How Alberta Can Achieve Fiscal Sustainability from Its Resource Revenues." C.D. Howe Institute Commentary, no. 263, May 2008. Web.

Shukin, Nicole. "Animal Capital: The Politics of Rendering." Diss. University of Alberta, 2005. Print.

———. "Place Minus People: the Math of Evacuation." Space + Memory = Place: Association for Literature, Environment and Culture in Canada (ALECC) 2012 Conference. University of British Columbia, Okanagan, Kelowna, BC. 9 Aug. 2012. Opening Plenary.

———. "The Trouble with Twitter: Protesting Twitter's Protest Value." *Briarpatch* Mar./Apr. 2014: 40. Print.

Silverstone, Peter. *World's Greenest Oil: Turning the Oil Sands from Black to Green*. Edmonton: PHS Holdings, 2010. Print.

Simpson, Zachary. "The Truths We Tell Ourselves: Foucault on *Parrhesia*." www.academia.edu, Mar. 2011. Web.

Sinnema, Jodie. "Nothing New in Celeb's Jabs, She Says." *Edmonton Journal* 18 Sept. 2013: A11. Print.

Sonntag, Jennifer. "Beyond Petroleum? Answers in the Gothic Heart of the Gulf." *on earth: A Survival Guide for the Planet*. Community Blog, 15 June 2010. Web.

Spivak, Gayatri Chakravorty. "Criticism, Feminism, and the Institution." *Intellectuals: Aesthetics, Politics, Academics*. Ed. Bruce Robbins. Minneapolis: U of Minnesota P, 1990. 153–72. Print.

———. *A Critique of Postcolonial Reason: Toward a History of the Vanishing Present*. Cambridge, MA: Harvard UP, 1999. Print.

Steiner, George. *After Babel: Aspects of Language and Translation*. 1975. New York: Oxford UP, 1998. Web.

Stenson, Fred. "Streamlined Away: The World's Fastest Trade Conquest." *Alberta Views* Dec. 2012: 18. Print.

Stephenson, Amanda. "Travel Alberta Demands Anti-Oilsands Film Trailer Be Yanked from YouTube." *Calgary Herald* 26 Aug. 2013. Web.

Stoekl, Allan. *Bataille's Peak: Energy, Religion, and Postsustainability*. Minneapolis: U of Minnesota P, 2007. Print.

———. "Unconventional Oil and the Gift of the Undulating Peak." *Imaginations: Journal of Cross-Cultural Image Studies* 3.2 (2012). Web.

Stolte, Elise. "Cree Lawsuit Would Drain Energy Royalties." *Edmonton Journal* 12 June 2009: B6. Print.

Strong, Tracy B. *Friedrich Nietzsche and the Politics of Transfiguration*. Berkeley: U of California P, 1975.

Sturgeon, Theodore. Introduction. *World Soul*. Trans. Antonina W. Bouis. New York: Macmillan, 1978.

Suncor Energy. *2009 Report on Sustainability*. Web. 2 July 2015.

———. "Discovery of the Ankylosaur at Suncor Energy's Millennium Mine." Online video. *YouTube*, 26 July 2011. Web.

"Sustainability." OED *Online*. Oxford UP, June 2015. Web.

Syncrude Canada Ltd. Public Affairs. *Are the Oil Sands Being Responsibly Developed?: 2010/11 Sustainability Report*. Fort McMurray, AB: Syncrude Canada Ltd., 2011. PDF file.

———. *Canadians Want Responsible Development of the Oil Sands: 2007 Sustainability Report*. Fort McMurray, AB: Syncrude Canada Ltd., 2007. PDF file.

———. *Focused: 2006 Sustainability Report*. Fort McMurray, AB: Syncrude Canada Ltd., 2006. PDF file.

———. *Synergy: 2008/09 Sustainability Report*. Fort McMurray, AB: Syncrude Canada Ltd., 2009. PDF file.

Szeman, Imre. "Introduction to Focus: Petrofictions." *American Book Review* 33.3 (2012): 3. Print.

———."System Failure: Oil, Futurity, and the Anticipation of Disaster." *South Atlantic Quarterly* 106.4 (2007): 805–23. Print.

Theresa. "How Suncor Saved an Ankylosaur." *McMurray Musings*, 27 July 2011. Web.

———. *McMurray Musings: Myth, Madness, and Making It*. Blog. Web. Various dates.

Thibaudeau, Colleen. "All My Nephews Have Gone to the Tar Sands." *Writing the Terrain: Travelling Through Alberta With the Poets*. Ed. Robert M. Stamp. Calgary: U of Calgary P, 2005. 248. Print.

Thibault, Benjamin. "Launch of Alberta's New Single Regulator Leaves Room for Improvement." www.pembina.org, 17 June 2013. Web.

Thomson, Graham. "The Grassroots Insistent on Green." *Edmonton Journal* 25 May 2014. Web.

Tietge, David J. "Rhetoric Is Not Bullshit." *Bullshit and Philosophy*. Ed. Gary L. Hardcastle and George A. Reisch. Chicago: Open Court, 2006. 229–40. Print.

Timoney, Kevin, and Peter Lee. "CNRL's Persistent 2013–2014 Bitumen Releases near Cold Lake, Alberta: Facts, Unanswered Questions, and Implications." Global Forest Watch. Web. 2 June 2014.

Tipping Point: The Age of the Oil Sands. Dir. Tom Radford and Niobe Thompson. Clearwater Documentary. *The Nature of Things*. CBC-TV. 7 Mar. 2013. Web.

Treaty 7 Elders and Tribal Council with Walter Hildebrandt, Dorothy First Rider, and Sarah Carter. *The True Spirit and Original Intent of Treaty 7*. Montreal and Kingston: McGill-Queen's UP, 1996. Print.

Treaty 8 Alberta Chiefs. "Treaty 8 Consultation Policy Position Paper." 30 Sept. 2010. www.treaty8.ca. Web.

Triandis, Harry C. *Fooling Ourselves: Self-Deception in Politics, Religion, and Terrorism*. Westport, CT: Praeger, 2009.

Turner, Chris. "Paradigm Shift: Enter the Second Industrial Age." *Alberta Views* July/Aug. 2011: 30–35. Print.

University of Alberta. "What's Next: Oil Sands Technology Could Save Millions of Lives." Advertisement. www.whatsnext.ualberta.ca, 13 Sept. 2013. Web.

"Untoward." *OED Online*. Oxford UP, June 2015. Web.

Van Camp, Richard. *Godless But Loyal to Heaven*. Winnipeg: Enfield & Wizenty, 2012.

Vanderklippe, Nathan. "Ambitious Plans for Oil Sands Would Create Lakes from Waste." *Globe and Mail* 2 Oct. 2012. Web.

Vermeulen, Pieter. "Community and Literary Experience in (between) Benedict Anderson and Jean-Luc Nancy." *Mosaic* 42.4 (2009): 95–111. Print.

Wardle, Elizabeth. "Identity, Authority, and Learning to Write in New Workplaces." *Enculturation* 5.2 (2004). Web.

Watt, Alison. "Creekwalker." *Poems for an Oil-Free Coast.* Victoria, BC: Red Tower Bookworks, 2012. Print.

Weber, Bob. "Excluded First Nations Head to Court." *Edmonton Journal* 12 June 2014: B3. Print.

Westman, Clinton N. "Cautionary Tales: Making and Breaking Community in the Oil Sands Region." *CJS: Canadian Journal of Sociology/Cahiers Canadiens de Sociologie* 38.2 (2013): 211–31. Print.

Weyler, Rex. Foreword. *The Enpipe Line: 70,000 km of Poetry Written in Resistance to the Northern Gateway Pipelines Proposal.* Smithers, BC: Creekstone P, 2012. Web.

Whaley, Susan. *Rudy Wiebe and His Works.* Toronto: ECW P, 1986.

Wiebe, Rudy. "The Angel of the Tar Sands." *The Angel of the Tar Sands and Other Stories.* Toronto: McClelland & Stewart, 1982. 188–91. Print.

———. "Passages by Land." *The Narrative Voice: Short Stories and Reflections by Canadian Authors.* Ed. John Metcalf. Toronto: McGraw-Hill Ryerson, 1972. 257–60. Print.

Wiebe, Rudy, and Theatre Passe Muraille. *Far as the Eye Can See.* Edmonton: NeWest P, 1977. Print.

Wingrove, Josh. "Nature Walks Jolt Hungry Predators, Study Says." *Globe and Mail* 23 Sept. 2011. Web.

———. "Syncrude to Pay $3 Million for Duck Deaths." *Globe and Mail* 22 Oct. 2010. Web.

Wohlberg, Meagan. "Fort McKay First Nation Objects to Oilsands Project." *Northern Journal* 29 Apr. 2013. Web.

Wong, Rita. "night gift." *The Enpipe Line: 70,000 km of Poetry Written in Resistance to the Northern Gateway Pipelines Proposal.* Smithers, BC: Creekstone P, 2012. Web.

———. "J28." *Decolonization: Indigeneity, Education, and Society.* https://decolonization.wordpress.com, 28 Jan. 2013. Web.

Yaeger, Patricia. "Editor's Column: The Death of Nature and the Apotheosis of Trash; or, Rubbish Ecology." *PMLA* 123.2 (2008): 321–39. Print.

———. "The Pipeline as Political Narrative." Petrocultures: Oil, Energy, and Culture. Edmonton, AB. 7 Sept. 2013. Conference Presentation.

Zantingh, Matthew. "When Things Act Up: Thing Theory, Actor-Network Theory, and Toxic Discourse in Rita Wong's Poetry." *Interdisciplinary Studies in Literature and the Environment* 20.3 (2013): 622–46. Web.

Žižek, Slavoj. *First as Tragedy, Then as Farce.* New York: Verso, 2009. Print.

———. *In Defense of Lost Causes.* New York: Verso, 2008. Print.

———. *Living in the End Times.* New York: Verso, 2010. Print.

———. "Use Your Illusion." *London Review of Books* 14 Nov. 2008. Web.

Zwicky, Jan. *Lyric Philosophy.* Toronto: U of Toronto P, 1992. Print.

———. Talk and Reading for the Environmental Research and Studies Centre. University of Alberta. 4 Oct. 2007. Lecture.

———. *Wisdom and Metaphor.* Kentville, NS: Gaspereau P, 2003. Print.

Index

Page numbers in bold indicate photos.